Introduction	i
1	1 Minute Calculus: X-Ray and Time-Lapse Vision	1
2	Practicing X-Ray and Time-Lapse Vision	5
3	Expanding Our Intuition	11
4	Learning The Official Terms	15
5	Music From The Machine	22
6	Improving Arithmetic And Algebra	25
7	Seeing How Lines Work	29
8	Playing With Squares	35
9	Working With Infinity	40
10	The Theory Of Derivatives	44
11	The Fundamental Theorem Of Calculus (FTOC)	48
12	The Basic Arithmetic Of Calculus	52
13	Finding Patterns In The Rules	58
14	The Fancy Arithmetic Of Calculus	62
15	Discovering Archimedes' Formulas	68
	Afterword	76
	Appendix: Learning Checklist	78
	Appendix: Calculus Study Plan	80

INTRODUCTION

Hi! It looks like you're interested in learning Calculus. I like you already.

This book isn't a collection of practice problems or formal theories. Hundreds of textbooks handle that quite well; this is the guide I wish they tucked into their front cover.

The goal is to help you:

- Grasp the essence of Calculus in hours, not months

- Develop lasting, practical insights you can apply to your own life

- Enjoy a subject often considered humorless

- Solve an end-to-end problem on your own

- Build an intuitive foundation for classroom study

Most Calculus courses force you to build the car before driving it. Shouldn't you master the physics of acceleration and the chemistry of gasoline before touching the wheel? (*Sure, if you want to kill someone's interest in cars.*)

I'm here to yank you from class, put you in a go-kart, and let you ride around Calculus Town. Yes, you'll take control of the wheel (try to avoid the pedestrians). Yes, you'll make a few mistakes without *perfect* knowledge of the internals. But you'll be having fun. True understanding comes from experiencing ideas yourself, not having them lectured to you.

After some practice, you may ignite the curiosity to explore the details regular textbooks offer. (*How can I go faster? Handle corners better? Which fuel works best?*)

A few minutes into Chapter 1, and you'll visualize what Calculus does. After an hour, you'll analyze concepts using the Calculus lingo and mindset. A few days later, after internalizing the ideas, you'll begin solving famous problems on your own.

So, how do we approach a notoriously difficult subject? Intuition-first.

Learning Strategy: Blurry To Sharp

What's a better learning strategy: covering a subject in full detail from top-to-bottom, or progressively sharpening a quick overview?

When we're ready, we "perform" math by doing the calculations ourselves, and diving into the detailed theory. But only if you want! Decide for *yourself* what level of understanding you'd like to reach.

Here's a guide to the book, depending on the level you'd like to reach:

- Intuitive Appreciation (Chapters 1-3)
- Technical Description (Chapters 4-5)
- Theory I (Chapters 6-8)
- Theory II (Chapters 9-14)
- Performance (Chapter 15)

The appendix includes a study plan if you wish to follow a formal course.

Learning Strategy: Study History

Calculus began when Archimedes realized shapes could be split into parts and rearranged. After a lifetime of effort, he discovered connections between spheres, circles and other shapes that were later etched onto his tomb.

Unfortunately, Calculus courses are taught out of order. The difficult, modern concepts are taught first (limits, developed 1800s) and the intuitive foundations are saved for the end (integrals, imagined 250 B.C.).

This book takes Archimedes' approach: learn to *see* what Calculus can do, then layer in the theory as needed. By then, your interest has been piqued and you can read a modern textbook for the details (see recommendations in the final chapter).

Please don't feel obligated to reach the Performance level of Calculus. For most of us (myself included), Appreciation and Description give practical, day-to-day insights. The first few chapters of this course are all that's needed to have a better intuition than I had despite dozens of engineering classes.

Sound good? Let's dive in.

Email Updates

BetterExplained.com provides high-quality, intuitive explanations to millions of readers each year. If you'd like updates on Calculus and related topics, sign up at:

`http://betterexplained.com/newsletter`

Book Webpage

Nobody likes typing URLs by hand. For clickable links for to the URLs in this book, notes, and other resources, visit:

`http://betterexplained.com/calculus/book`

After a single class, which strategy gives you a better understanding of the material? Which helps you predict how later parts fit together? Which is more *fun*?

The linear, official, approach doesn't work for me. Starting with a rough outline and gradually improving it keeps our interest and helps us see how the individual details are connected.

Most textbooks take the top-down approach, while this book is a blurry-to-sharp guide. The fine details are out there when you need them.

Learning Strategy: Appreciation To Performance

Next question: Do you need to be a musician to enjoy a song? No way.

There are several levels of music understanding:

- **Intuitive Appreciation:** Just enjoying the music.
- **Natural Description:** Humming a tune you heard or made up.
- **Symbolic Description:** Reading and writing the sheet music.
- **Theory:** Explaining how harmonies work, why minor scales are somber, etc.
- **Performance:** Playing the official instruments.

Math knowledge is the same. Start by appreciating, even enjoying, the idea. Describe it with your own words, in English. Then, learn the official symbols to make communication easier ("$2 + 3 = 5$" is better than "Two plus three equals five", right?).

The Legal Stuff

All content is copyright (c) 2015 Kalid Azad, except the following images used under Wikipedia's Creative Commons license:

- "Portrait of Isaac Newton" by Sir Godfrey Kneller `http://commons.wikimedia.org/wiki/File:GodfreyKneller-IsaacNewton-1689.jpg`

- "Tree Rings" by Arnoldius `http://commons.wikimedia.org/wiki/File:Tree_rings.jpg`

- "Acropoclipse" by Elias Politis `http://commons.wikimedia.org/wiki/File:Acropoclipse.jpg`

- "Geometric Series" by Indolences `http://commons.wikimedia.org/wiki/File:Geometric_series_14_square2.svg`

You may use any of this material for educational use with attribution; for commercial use, please contact me using `http://betterexplained.com/contact`.

CHAPTER 1

1 Minute Calculus: X-Ray and Time-Lapse Vision

We usually take shapes, formulas, and situations at face value. Calculus gives us two superpowers to dig deeper:

- **X-Ray Vision:** You see the hidden pieces inside a pattern. You don't just see the tree, you know it's made of rings, with another growing as we speak.

- **Time-Lapse Vision:** You see the future path of an object laid out before you (cool, right?). "Hey, there's the moon. For the next few days it'll be white, but on the sixth it'll be low in the sky, in a color I like. I'll take a photo then."

1

CHAPTER 1. 1 MINUTE CALCULUS: X-RAY AND TIME-LAPSE VISION

So how is Calculus useful? Well, just imagine having X-Ray or Time-Lapse vision to use at will. For an object or scenario we care about, how was it put together? What will happen to it down the line?

(Strangely, my letters to Marvel about Calculus-man have been ignored to date.)

1.1 Calculus In 10 Minutes: See Patterns Step-By-Step

What do X-Ray and Time-Lapse vision have in common? They examine patterns step-by-step. An X-Ray shows the individual slices inside, and a Time Lapse puts each future state next to the other.

This seems pretty abstract. Let's look at the equations for circumference, area, surface area, and volume:

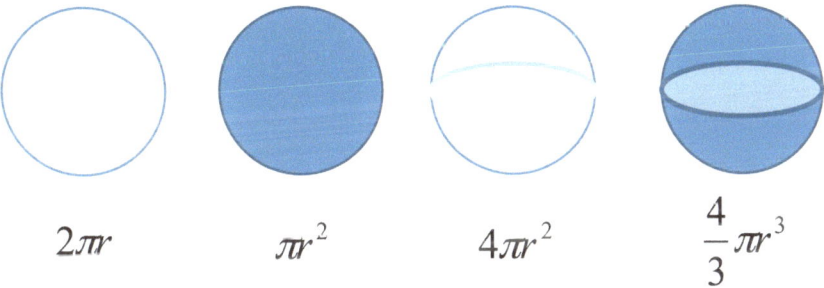

We have a vague feeling these formulas are connected, right?

Let's turn on our X-Ray vision and see where this leads. Suppose we know *circumference* = $2\pi r$ and we want to figure out the equation for area. What can we do?

This is a tough question. Squares are easy to measure, but how do we work out the size of an ever-curving shape?

Calculus to the rescue. Let's use our X-Ray vision to realize a disc is really just a bunch of rings put together. Similar to a tree trunk, here's a "step-by-step" view of a circle's area:

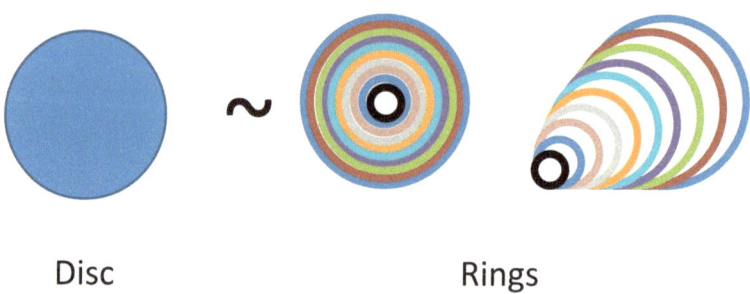

How does this viewpoint help? Well, let's unroll those curled-up rings into straight lines, so they're easier to measure:

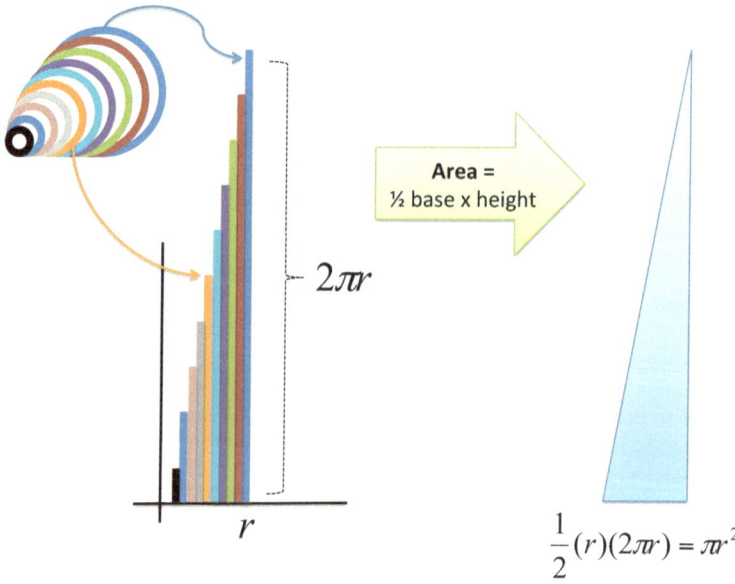

$$\frac{1}{2}(r)(2\pi r) = \pi r^2$$

Whoa! We have a bunch of straightened-out rings that form a triangle, which is much easier to measure. (Wikipedia has an animation[1].)

The height of each ring depends on its original distance from the center; the ring 3 inches from the center would have a height of $2\pi \cdot 3$ inches. The smallest ring is a pinpoint, more or less, without any height at all. The height of the largest ring is the full circumference ($2\pi r$).

And because triangles are easier to measure than circles, finding the area isn't too much trouble. The area of the "ring triangle" = $\frac{1}{2}base \cdot height = \frac{1}{2}r(2\pi r) = \pi r^2$, which is the formula for a circle's area!

Our X-Ray vision revealed a simple, easy-to-measure structure within a curvy shape. We realized a circle and a set of glued-together rings were really the same. From another perspective, a filled-in disc is really just the "time lapse" of a single ring that grew larger.

1.2 So... What Can I Do With Calculus?

Remember learning arithmetic? You learned how to count out a number, and how to combine it with others (add/subtract, multiply/divide, take exponents/roots). Technically, counting isn't necessary, as our caveman ancestors did "fine" (survived) without it.

But, having a specific notion of quantity makes navigating the world easier. You don't have a "big" and "small" pile of rocks: you have an exact count. You know how many arrows can be given to each hunter, or whether the gathered berries are enough for the tribe.

Even better, arithmetic gives us metaphors that go beyond strict calculations. It has sharpened our descriptions of everything, letting us clarify everything from spiciness levels and movie ratings (1 to 5) to our mood (1 to 10). Specific measurements are a useful idea, and hard to give up once seen.

Calculus trains us in two new metaphors: splitting apart and gluing together. A pattern can be separated into parts, and the parts can be progressively combined into the full pattern.

Is this viewpoint necessary for survival? Nope. But it is interesting.

Numbers and equations describe what we *have*, but Calculus explains the steps that *got us there*. Instead of just the cookie, we can see the recipe.

Sure, Calculus appears in science because a step-by-step blueprint is more useful than being handed a final result. But in everyday scenarios, we have a nice perspective to turn on: What steps got us here? Are there any pros or cons to that approach? And based on these steps, where are we going next?

Let's feel what a Calculus perspective is like.

---

[1] Visit http://betterexplained.com/calculus/book for clickable links to extra resources.

CHAPTER 2

PRACTICING X-RAY AND TIME-LAPSE VISION

Calculus trains us to use X-Ray and Time-Lapse vision, such as re-arranging a circle into the "ring triangle" we saw in the previous chapter. This makes finding the area... well, if not exactly *easy*, much more manageable.

But we were a little presumptuous. Must *every* circle in the universe be made from rings?

Heck no! We're more creative than that. Here's a few more ways we could have X-Rayed the circle:

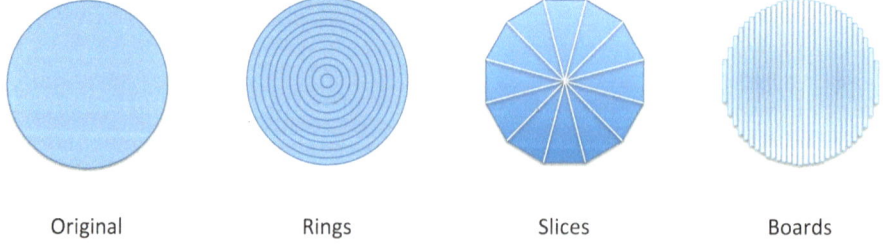

Original   Rings   Slices   Boards

We could imagine a circle as a set of rings, pizza slices, or vertical boards. Each blueprint is a different step-by-step strategy in action.

2.1 Ring-by-ring Analysis

Using your Time-Lapse vision, imagine how the ring-by-ring strategy accumulates over time:

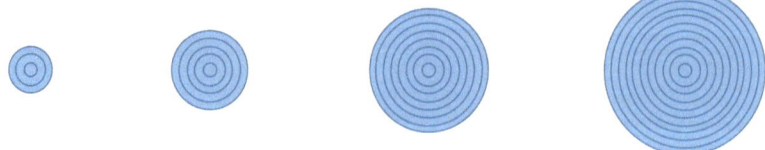

What's interesting about a ring-by-ring progression?

- Each intermediate stage is an entire "mini circle" on its own. i.e., when we're halfway done, we still have a circle, just one with half the regular radius.

- Each step is an increasing amount of work. Just imagine plowing a circular field and spreading the work over several days. On the first day, you start at the center and don't even move. The next, you make take the tightest turn you can. Then you start doing laps, larger and larger, until you are circling the entire yard on the last day.

- The work is reasonably predictable, which may help planning. If we know it's an extra minute for each lap, then the 20th ring will take 20 minutes.

- Most of the work happens in the final laps. In the first 25% of the timelapse, we've barely grown: we're adding tiny rings. Near the end, we start to pick up steam by adding long rings, each nearly the final size.

Now let's get practical: why do trees follow a ring pattern?

A big tree must grow from a complete smaller tree. With the ring-by-ring strategy, we're always building on a complete, fully-formed circle. We aren't trying to grow the "left half" of the tree and then work on the right side.

In fact, many natural processes that grow (trees, bones, bubbles, etc.) take this inside-out approach.

2.2 Slice-by-slice Analysis

Now think about a slice-by-slice progression. What do you notice?

- We contribute the same amount with each step. Even better, the parts are identical. This may not matter for math, but in the real world (e.g., cutting a cake), we want to use the same motion when cutting each slice.

- Because the slices are symmetrical, we can use shortcuts like making cuts across the entire shape. These "assembly line" speedups work well when generating identical components.

- Progress is extremely easy to measure. If we have 10 slices, then at slice 6 we are exactly 60% done (by both area and circumference).

- We follow a sweeping circular path, never retracing our steps from an "angular" point of view. When carving out the rings, each step involved the full 360 degrees.

Let's think about the real world: what objects use the slice-by-slice pattern, and why?

Well food, for one. Cake, pizza, pie: we want everyone to have an equal share. Slices are simple to cut, we get nice speedups (cutting across the cake), and it's easy to see how much is remaining. Imagine cutting circular rings from a pie and estimating how much is left.

Now think about radar scanners: they sweep a beam in a circle, "clearing" a slice of sky before moving to another angle. This strategy does leave a blind spot in the angle you haven't yet covered, a tradeoff you're hopefully aware of.

Contrast this to sonar used by a submarine or bat, which sends a sound "ring" propagating in every direction. That works best for close targets (covering every direction at once). The drawback is that unfocused propagation gets much weaker the further out you go, as the initial energy is spread out over a larger ring. We use megaphones and antennas to focus our signals into beams (thin slices) to get the most range for our energy.

Logistically, if we're building a circular shape from a set of slices (like the folded sections of a paper fan), it helps to have every part be identical. Figure out the best way to make a single slice, then mass produce them. Even better: if one part can collapse, the entire shape can fold up!

2.3 Board-by-board Analysis

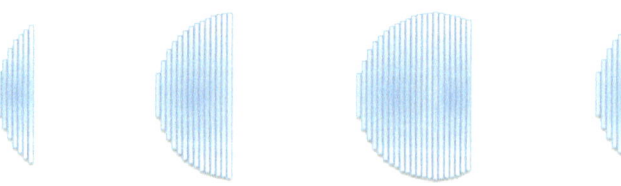

Getting the hang of X-Rays and Time-lapses? Great. Look at the progression above, and spend a few seconds thinking of the pros and cons. Don't worry, I'll wait.

Ready? Ok. Here's a few of my observations:

- This is a very robotic pattern, with boards neatly arranged left-to-right.

- The contribution from each step starts small, gradually gets larger, maxes out in the middle, and begins shrinking again.

- Our progress is somewhat unpredictable. Sure, at the halfway mark we've finished half the circle, but the pattern rises and falls which makes it difficult to analyze. By contrast, the ring-by-ring pattern changed the same amount each time, always increasing. It was clear that the later rings would add the most work. Here, it's the middle section which seems to be doing the heavy lifting.

Ok, time to figure out where this pattern shows up in the real world.

Decks and wooden structures, for one. When putting down wooden planks, we don't want to retrace our steps (especially if there are other steps involved, like painting). Just like a tree needs a fully-formed circle at each step, a deck insists upon using the linear boards we can find at Home Depot.

In fact, any process with a strict "pipeline" might use this approach: finish a section and move onto the next. Think about a printer that has to spray a pattern as the paper is fed through (or these days, a 3d printer). The printer sees a position only once, so it better make it count!

The circle doesn't need to be a literal shape. It could represent a goal you're trying to accomplish, whether an exercise plan or topics to cover in a counseling session.

The board approach suggests you start small, work your way up, then ease back down. The pizza-slice approach could be tolerable (identical progress every day), but rings could be demoralizing: every step requires more effort than the one before.

2.4 Getting Organized

So far, we've been using natural descriptions to explain our thoughts: "Take a bunch of rings" or "Cut the circle into pizza slices". This conveys a general notion, but it's a bit like describing a song with "Dum-de-dum-dum" – you're probably the only one who knows what you mean. But a little organization can make your intent perfectly clear.

To start, we can explain how we're splitting the shape into steps. I like to imagine a little arrow in the direction we move as we cut out each piece:

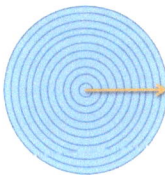

In my head, I'm moving along the yellow line, calling out the steps as we go (step forward, make a ring, step forward, make a ring...).

And while the arrow shows how the rings are made, the steps are hard to visualize because they're jammed inside the circle. As we saw in the first lesson, we can pull out the individual steps and line them up:

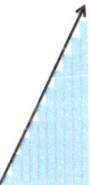

We draw a black arrow to show the trend in the size of each step. Pretty nice, right? We can tell, at a glance, that the rings are increasing, and by the same amount each time (the trend line is straight, like a set of stairs).

Math fans and neurotics both enjoying lining up the pieces. There is something soothing about it, I suppose: who doesn't want to line up all the pencils that are scattered on the floor?

And since we're on the topic, we might as well organize the other patterns too:

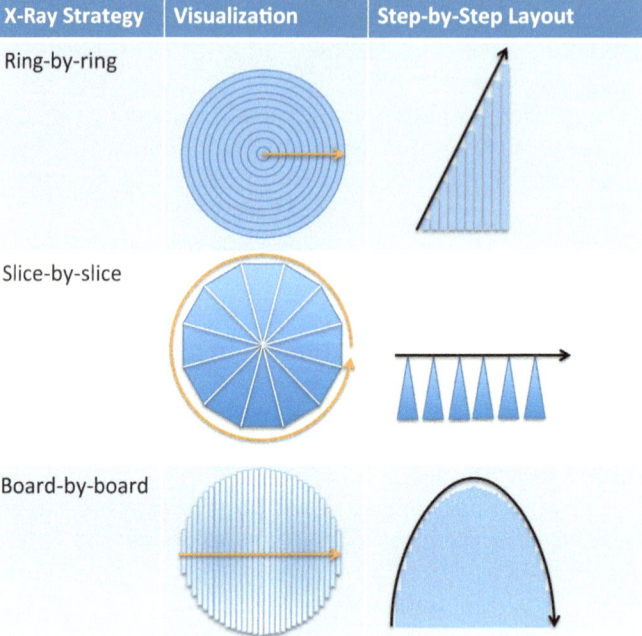

Ah! Now it's much easier to compare each X-Ray strategy:

- With circular rings, steps increase steadily (upward sloping line)

- With triangular slices, steps stay the same size (flat, horizontal line)

- With rectangular boards, steps get larger, peak, then get smaller (up and down). The trend line looks elongated because the boards have been pushed down to line up at the bottom.

These charts make the strategy comparisons easier, wouldn't you say? Sure. But wait, isn't that trendline looking like a dreaded x-y graph?

Yep. Unfortunately, many classes simply present the black trend line, without showing you the pieces it represents. That's a recipe for pain: be explicit about what the graph means!

The black trend line is the super-summarized description of an X-Ray strategy. We're showing the size of each piece (the graph height) and how their size is changing (trend direction).

The distinction between a ring, slice and board isn't important: in Calculus, they're all pieces of the whole pattern. Words like "slice", "ring" or "board", are just descriptive versions of "piece of the whole", and are otherwise interchangeable.

In this guide we'll keep graphs to the level seen above: trend lines, with the individual pieces shown. This is a foundation for later, performance-based classes where you may work with graphs directly. But just for reference, Archimedes laid the foundations of Calculus without x-y graphs, and found his results without them.

2.5 Questions

Are things starting to click a bit? Thinking better with X-Rays and Time-lapses?

1. Can you describe a grandma-friendly version of what you've learned?

2. Let's expand our thinking into the 3rd dimension. Can you think of a few ways to build a sphere? (No formulas, plain-English descriptions are fine.)

PS. It may bother you that our steps create a "circle-like" shape, but not a *real, smooth* circle. Great question, and we'll get to that. But to be fair, it must also bother you that the square pixels on a computer screen make "letter-like" patterns, but not *real, smooth* letters. And somehow, the "letter-like patterns" convey the same information as real letters!

CHAPTER 3

EXPANDING OUR INTUITION

I hope you thought about the question from last time: how do we take our X-Ray strategies into the 3rd dimension?

Here's my take:

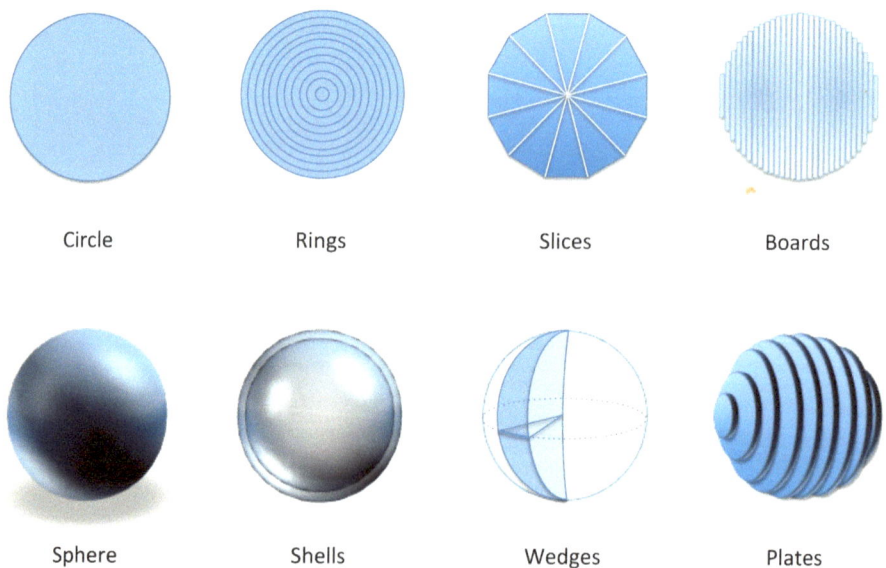

- Rings become *shells*, a thick candy coating on a delicious gobstopper. Each layer is slightly bigger than the one before.

- Slices become *wedges*, identical sections like slices of an orange.

- Boards become *plates*, thick discs which can be stacked together. (I sometimes daydream of opening a bed & breakfast that only serves spherical stacks of pancakes.)

The 3d segments can be seen as being made from their 2d counterparts. For example, we can spin an individual ring like a coin to create a shell. A wedge looks like a bunch of pizza slices (of different sizes) stacked on top of each other.

11

Lastly, we can imagine spinning a board to make a plate, like carving a wooden sphere with a lathe (video).

The tradeoffs in 3d are similar to the 2d versions:

- Organic processes grow in shell-by-shell layers (pearls in an oyster).

- Fair divisions require wedges (cutting an apple for friends).

- The robotic plate approach seems easy to manufacture (weightlifting plates).

An orange is an interesting hybrid: from the outside, it appears to be made from shells, growing over time. But on the inside, it forms a symmetric internal structure with wedges – good for evenly distributing seeds, right? We could analyze it both ways.

3.1 Exploring The 3d Perspective

In the first lesson we had the vague notion that the circle/sphere formulas were related:

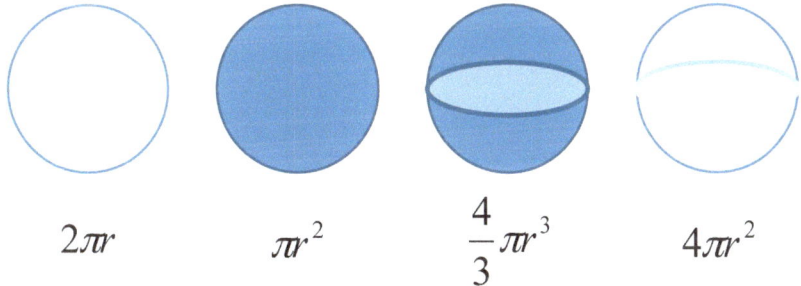

$2\pi r$ \qquad πr^2 \qquad $\frac{4}{3}\pi r^3$ \qquad $4\pi r^2$

With our X-Ray and Time-Lapse skills, we have a specific idea for how:

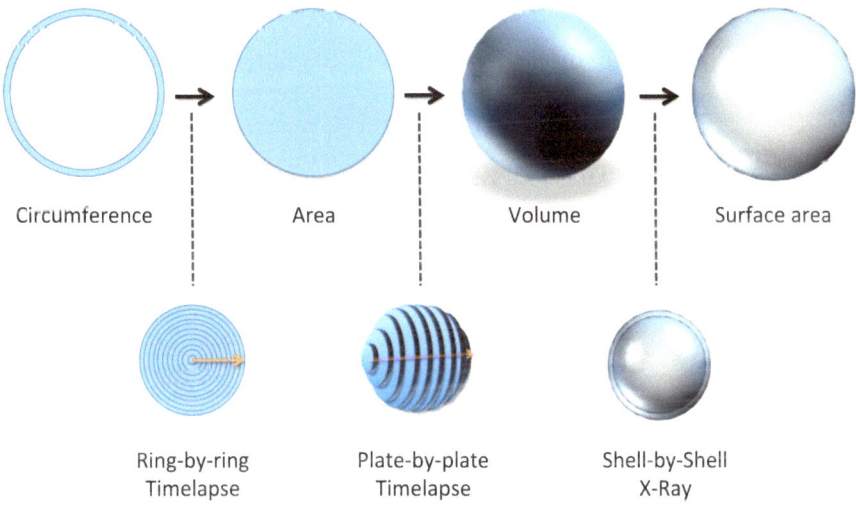

- **Circumference**: Start with a single ring.
- **Area:** Make a filled-in disc with a ring-by-ring time lapse.
- **Volume:** Make the circle into a plate, and do a plate-by-plate time lapse to build a sphere.
- **Surface area:** X-Ray the sphere into a bunch of shells; the outer shell is the surface area.

Wow! We now have detailed descriptions of how one formula is related to the other. We know, intuitively, how to morph shapes into alternate versions by thinking "Time-Lapse this" or "X-Ray that". We could even work backwards: starting with a sphere, we can X-Ray it into plates, and then take one plate and X-Ray it into rings.

3.2 The Need For Math Notation

You might have noticed it's getting harder to describe your ideas. We're reaching for physical analogies (rings, boards, wedges) to explain our plans: "Ok, take that circular area, and try to make some discs out of it. Yeah, like that. Now line those discs up into the shape of a sphere...".

I love diagrams and analogies, but should they be *required* to explain an idea? Probably not.

Take a look at how numbers developed. At first, we used very literal symbols for counting: I, II, III, and so on. Eventually, we realized a symbol like V could take the place of IIIII, and even better, every digit can have its own symbol. (The number "1" reminds us of our line-based history.)

Math notation helped in a few ways:

- **Symbols are shorter than words.** Isn't "2 + 3 = 5" better than "two added to three is equal to five"? Fun fact: In 1557, Robert Recorde invented the equals sign, written with two parallel lines (=), because "noe 2 thynges, can be moare equalle". (*I agryee!*)

- **The rules do the work for us.** With Roman numerals, we're essentially recreating numbers by hand (why should VIII take so much effort to write compared to I? Just because 8 is larger than 1? Not a good reason!). Decimals help us "do the work" of expressing numbers, and make them easy to manipulate. So far, we've been doing the work of calculus ourselves: cutting a circle into rings, realizing we can unroll them, looking up the equation for area and measuring the resulting triangle. Couldn't the rules help us here? You bet. We just need to figure them out.

- **We generalized our thinking.** "2 + 3 = 5" is really "twoness + threeness = fiveness". It sounds weird, but we have an abstract quantity (not people, or money, or cows... just "twoness") and we see how it's related to other quantities. The rules of arithmetic are general-purpose, and it's our job to apply them to a specific scenario.

This last point is important. When learning addition, your teacher may have used literal apples to show that two plus three was five. With enough practice, you started using abstract symbols without needing a physical example, and "2 + 3 = 5" made sense.

Calculus is similar: it works on abstract equations like $f(x) = x^2$, but physical examples are a nice starting point. When we see a shape like this:

we can actually *see* what Calculus does as we apply a technique, instead of pushing symbols around. Eventually, we can convert the shape into its corresponding equation and work with symbols directly.

So, don't think Calculus requires a real-world object, any more than addition requires apples. It can analyze any shape or formula (a physics equation, business scenario, graph of a function) – shapes are just easier to start with.

CHAPTER 4

LEARNING THE OFFICIAL TERMS

We've been able to describe our thinking process with analogies (X-Rays, Time-Lapses) and diagrams:

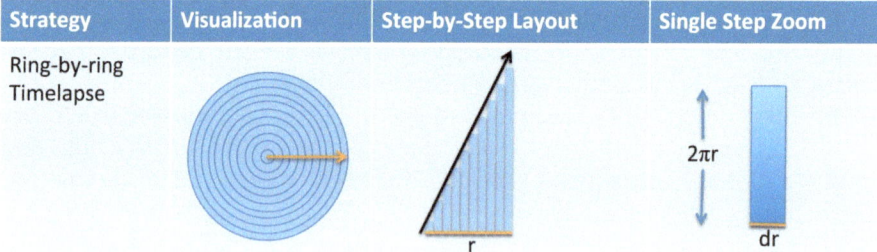

Strategy	Visualization	Step-by-Step Layout	Single Step Zoom
Ring-by-ring Timelapse			

However, this is a very elaborate way to communicate. Here's the Official Math® terms that describe our intuitive concepts:

Intuitive Concept	Formal Name	Symbol
X-Ray (split apart)	Take the derivative (derive)	$\frac{d}{dr}$
Time-lapse (glue together)	Take the integral (integrate)	$\int$
Arrow direction	Integrate or derive "with respect to" a variable.	dr implies moving along r
Arrow start/stop	Bounds or range of integration	$\int_{start}^{end}$
Slice	Integrand (shape being glued together, such as a ring)	Equation, such as $2\pi r$

Let's walk through the fancy names.

15

4.1 The Derivative

The **derivative** is the pattern of slices we get as we X-Ray a shape. It's indicated by the black trend line, which might be flat, rising constantly, rising and falling, etc. Now here's the trick: although the derivative generates the entire sequence of slices, we can also extract a *single* slice.

Think about a function like $f(x) = x^2$. It describes every possible squared value (1, 4, 9, 16, 25, etc.), and we can graph them all on a chart. But, we can also ask for the value of $f(x)$ at a specific value, such as at $x = 3$.

The derivative is similar. Officially, it's the complete pattern of slices we get after X-Raying a shape. However, we can pull out an individual slice by asking for the derivative *at* a certain value. (The derivative is a function, just like $f(x) = x^2$, and mathematicians assume you're talking about the entire function unless you ask for a specific slice.)

So, what do we need to find the derivative? Just the shape to split apart, and the path to follow as we cut it up (the orange arrow). The terminology is "derive <some pattern> with respect to <some direction>". For example:

- The derivative of a circle *with respect to* the radius creates rings (which always increase)

- The derivative of a circle *with respect to* the perimeter creates slices (which are equal-sized)

- The derivative of a circle *with respect to* the x-axis creates boards (which get larger, peak, and get smaller)

I agree that "with respect to" sounds formal: *Honorable Grand Poombah radius, it is with respect to you that we derive.* Math is a gentleman's game, I suppose.

Taking the derivative is also called "differentiating", because we are finding the difference between successive positions as a shape grows. As we grow the radius of a circle, the outer ring is the difference between the size of the current disc and the next size up.

4.2 The Integral, Arrows, and Slices

The **integral** is gluing together (Time-Lapsing) a bunch of slices and measuring the final result. For example, we glued together the rings (into a "ring triangle") and saw it accumulated to πr^2, aka the area of a circle.

Here's what we need to find the integral:

- **Which direction are we gluing the steps together?** Along the orange line (the radius, in this case)

- **When do we start and stop?** At the start and end of the arrow (we start at 0, no radius, and move to r, the full radius)

- **How big is each step?** Well... each item is a "ring". Isn't that enough?

CHAPTER 4. LEARNING THE OFFICIAL TERMS

Nope! We need to be specific. We've been saying we cut a circle into "rings" or "pizza slices" or "boards". But that's not specific enough; it's like a BBQ recipe that says "Cook meat. Flavor to taste."

Maybe an expert knows what to do, but we need more specifics. How large, exactly, is each step (technically called the "integrand")?

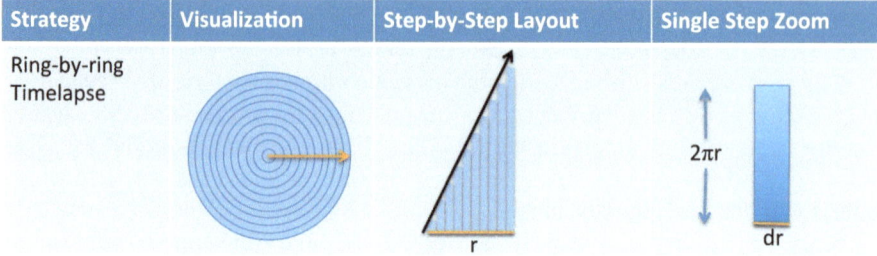

Ah. A few notes about the variables:

- If we are moving along the radius r, then dr is the little chunk of radius in the current step

- The height of the ring is the circumference, or $2\pi r$

There's several gotchas to keep in mind.

First, dr is its own variable, and not "d times r". It represents the tiny section of the radius present in the current step. This symbol (dr, dx, etc.) is often separated from the integrand by just a space, and it's assumed to be multiplied (written $2\pi r\, dr$).

Next, if r is the only variable used in the integral, then dr is assumed to be there. So if you see $\int 2\pi r$ this still implies we're doing the full $\int 2\pi r\, dr$. (Again, if there are two variables involved, like radius and perimeter, you need to clarify which step we're using: dr or dp?)

Last, remember that r (the radius) changes as we Time-Lapse, starting at 0 and eventually reaching its final value. When we see r in the *context of a step*, it means "the size of the radius at the current step" and not the final value it may ultimately have.

These issues are extremely confusing. I'd prefer we use r_{dr} for "r at the current step" instead of a general-purpose r that's easily confused with the max value of the radius. We can't change the symbols at this point, unfortunately.

4.3 Practicing The Lingo

Let's learn to speak like calculus natives. Here's how we can describe our X-Ray strategies:

Intuitive Visualization	Formal Description	Symbol
	derive the area of a circle with respect to the radius	$\frac{d}{dr}Area$
	derive the area of a circle with respect to the perimeter	$\frac{d}{dp}Area$
	derive the area of a circle with respect to the x-axis	$\frac{d}{dx}Area$

Remember, the derivative just splits the shape into (hopefully) easy-to-measure steps, such as rings of size $2\pi r \ dr$. We broke apart our lego set and have pieces scattered on the floor. We still need an integral to glue the parts together and measure the new size. The two commands are a tag team:

- The derivative says: "Ok, I split the shape apart for you. It looks like a bunch of pieces $2\pi r$ tall and dr wide."

- The integral says: "Oh, those pieces resemble a triangle – I can measure that! The total area of that triangle is $\frac{1}{2}base \cdot height$, which works out to πr^2 in this case."

Here's how we'd write the integrals to measure the steps we've made:

CHAPTER 4. LEARNING THE OFFICIAL TERMS

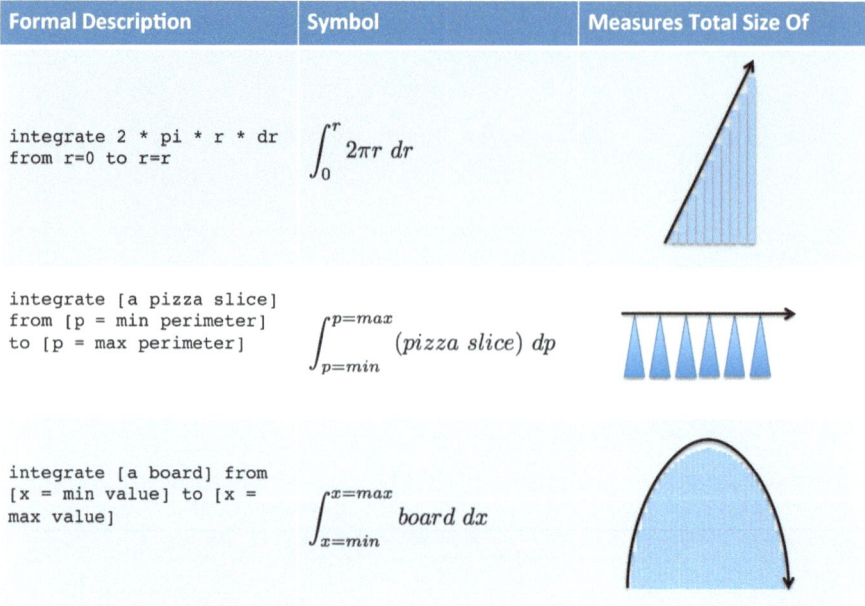

Formal Description	Symbol	Measures Total Size Of
integrate 2 * pi * r * dr from r=0 to r=r	$\int_0^r 2\pi r \, dr$	
integrate [a pizza slice] from [p = min perimeter] to [p = max perimeter]	$\int_{p=min}^{p=max} (pizza\ slice) \, dp$	
integrate [a board] from [x = min value] to [x = max value]	$\int_{x=min}^{x=max} board \, dx$	

A few notes:

- Often, we write an integrand as an unspecified "pizza slice" or "board" (use a formal-sounding name like $s(p)$ or $b(x)$ if you like). First, we setup the integral, and then we worry about the exact formula for a board or slice.

- Because each integral represents slices from our original circle, we know they will be the same. Gluing any set of slices should always return the total area, right?

- The integral is often described as "the area under the curve". It's accurate, but shortsighted. Yes, we are gluing together the rectangular slices under the curve. But this completely overlooks the preceding X-Ray and Time-Lapse thinking. Why are we dealing with a set of slices vs. a curve in the first place? Most likely, because those slices are easier than analyzing the shape itself (how do you "directly" measure a circle?).

4.4 Questions

At a high level, can you find another activity made *easier* with symbols, instead of using full English sentences? Would practitioners ever go back to written descriptions?

Math is just like that. Let's try a few phrases, even if we aren't fluent yet.

Question 1: Can you describe the integrals below in "Math English"?

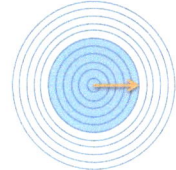 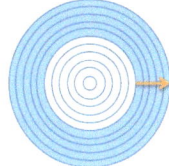

Assume the arrow spans half the radius. The description should follow the format:

```
integrate [size of step] from [start] to [end] with respect to [path variable]
```

Have an idea? Here's the answer for the first integral[1] and the second integral[2]. These links go to Wolfram Alpha, an online math solver, which we'll learn to use.

Question 2: Can you find the "Math English" way to describe our pizza-slice idea?

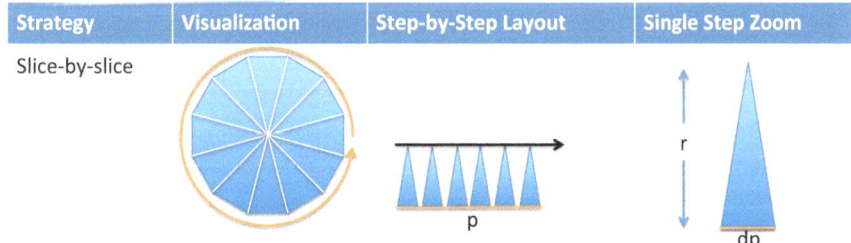

The math description should be something like this:

```
integrate [size of step] from [start] to [end] with respect to [path variable]
```

Remember that each slice is basically a triangle (so what's the area?). The slices move around the perimeter (where does it start and stop?). Have a guess for the command? Here it is, the slice-by-slice description[3].

Question 3: Can you describe how to move from volume to surface area?

---

[1] integrate 2 * pi * r * dr from r=0 to r=0.5r
[2] integrate 2 * pi * r * dr from r=0.5 to r=r
[3] integrate 1/2 * r * dp from p=0 to p=2*pi*r

CHAPTER 4. LEARNING THE OFFICIAL TERMS   21

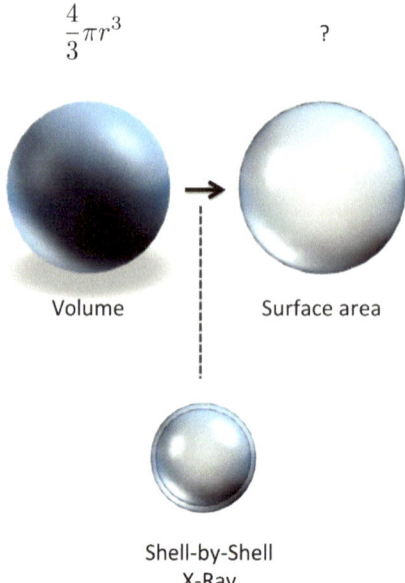

Assume we know the volume of a sphere is 4/3 * pi * r^3. Think about the instructions to separate that volume into a sequence of shells. Which variable are we moving through?

derive [equation] with respect to [path variable]

Have a guess? Great. Here's the command to turn volume into surface area[4].

---

[4]derive 4/3 pi * r^3 with respect to r

Chapter 5

MUSIC FROM THE MACHINE

In the previous lessons we've gradually sharpened our intuition:

- **Appreciation:** I think it's possible to split up a circle to measure its area
- **Natural Description:** Split the circle into rings from the center outwards, like so:

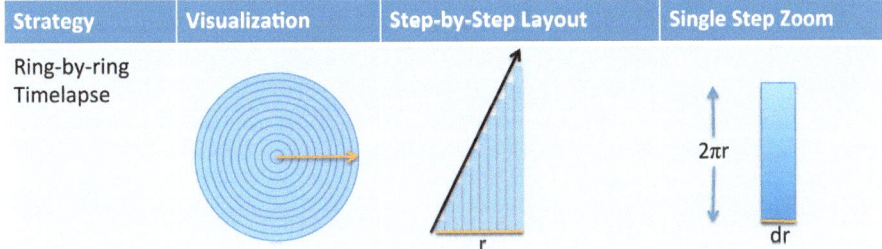

- **Formal Description**: integrate 2 * pi * r * dr from r=0 to r=r
- **Performance**: (*Sigh*) I guess I'll have to start measuring the area...

Wait! Our formal description is precise enough that a computer can do the work for us:

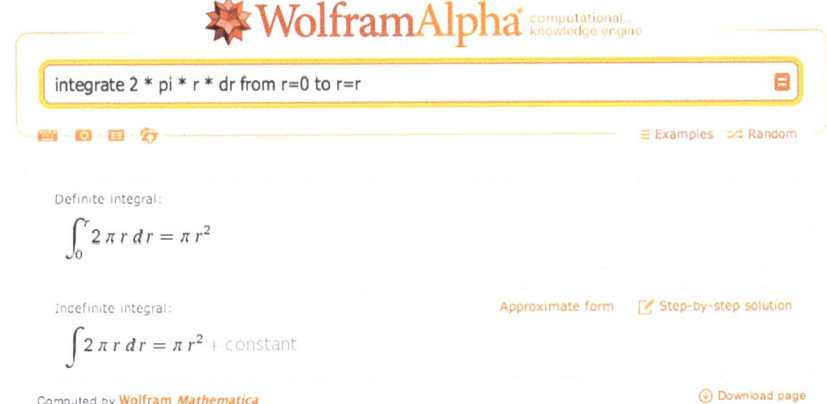

22

Whoa! We described our thoughts well enough that a computer did the legwork.

We didn't need to manually unroll the rings, draw the triangle, and find the area (which isn't overly tough in this case, but could have been). We saw what the steps would be, wrote them down, and fed them to a computer: boomshakalaka, we have the result. (Just worry about the "definite integral" portion for now.)

Now, how about derivatives, X-Raying a pattern into steps? Well, we can ask for that too:

Similar to above, the computer X-Rayed the formula for area and split it step-by-step as it moved. The result is $2\pi r$, the height of the ring at every position.

5.1 Seeing The Language In Action

Wolfram Alpha is an easy-to-use tool: the general format for calculus questions is

- `integrate [equation] from [variable=start] to [variable=end]`
- `derive [equation] with respect to [variable]`

That's a little wordy. These shortcuts are closer to the math symbols:

- `\int [equation] dr` - integrate equation (by default, assume we go from $r = 0$ to $r = r$, the max value)
- `d/dr equation` - derive equation with respect to r
- There's shortcuts for exponents (3^2 = 9), multiplication (3 * r), and roots (`sqrt(9) = 3`)

Now that we have the machine handy, let's try a few of the results we've seen so far:

Computer Description	Math Notation	Intuitive Notion
integrate 2 * pi * r * dr from r=0 to r=r	$\int_0^r 2\pi r\, dr$	
integrate 1/2 * r * dp from p=0 to p=2*pi*r	$\int_0^{2\pi r} \frac{1}{2} r\, dp$	
integrate 2 * sqrt(r^2 - x^2) from x=-r to x=r	$\int_{-r}^{r} 2\sqrt{r^2 - x^2}\, dx$	

Click the formal description to see the computer crunch the numbers. As you might have expected, they all result in the familiar equation for area. A few notes:

- The size of the wedge is $\frac{1}{2} base \cdot height$. The base is dp (the tiny section of perimeter) and the height is r, the distance from the perimeter back to the center.

- The size of the board is tricky. In terms of x & y coordinates, we have $x^2 + y^2 = r^2$, by the Pythagorean Theorem:

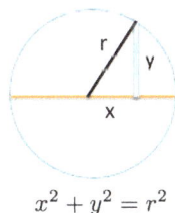

$$x^2 + y^2 = r^2$$

We solve for the height to get $y = \sqrt{r^2 - x^2}$. We actually need 2 copies of height, because y is the positive distance above the axis, and the board extends above and below. The boards are harder to work with, and it's not just you: Wolfram Alpha takes longer to compute this integral than the others!

The approach so far has been to immerse you in calculus thinking, and gradually introduce the notation. Some of it may be a whirl – which is completely expected. You're sitting at a cafe, overhearing conversation in a foreign language.

Now that you have the sound in your head, we'll begin to explore the details piece-by-piece.

CHAPTER 6

IMPROVING ARITHMETIC AND ALGEBRA

We've intuitively seen how calculus dissects problems with a step-by-step viewpoint. Now that we have the official symbols, let's see how to bring arithmetic and algebra to the next level.

6.1 Better Multiplication And Division

Multiplication makes addition easier. Instead of grinding through questions like 2 + 2 + 2 + 2 + 2 + 2 + 2 + 2 + 2 + 2 + 2 + 2 + 2, we can rewrite it as: 2 × 13.

Boomshakalaka. If you wanted 13 copies of a number, just write it like that!

Multiplication makes repeated addition easier. But there's a big limitation: we must use identical, average-sized pieces.

- What's 2 × 13? It's 13 copies of *the same element*.

- What's 100 / 5? It's 100 split into 5 *equal parts*.

Identical parts are fine for textbook scenarios, where you drive an unwavering 30mph for exactly 3 hours. The real world isn't so smooth. Calculus lets us accumulate or separate shapes according to their *actual*, not average, amount:

- **The derivative is a better type of division** that splits a shape along a path (into possibly different-sized slices)

- **The integral is a better type of multiplication** that accumulates a sequence of steps (which could be different sizes)

Operation	Example	Notes
Division	$\frac{y}{x}$	Split whole into identical parts
Differentiation	$\frac{d}{dx}y$	Split whole into (possibly different) parts
Multiplication	$y \cdot x$	Accumulate identical steps
Integration	$\int y \, dx$	Accumulate (possibly different) steps

25

Let's analyze our circle-to-ring example again. How does arithmetic/algebra compare to calculus?

Operation	Formula	Diagram
Division	$average\ step = \dfrac{Area}{radius} = \dfrac{\pi r^2}{r} = \pi r$	
Differentiation	$actual\ steps = \dfrac{d}{dr}\pi r^2 = 2\pi r$	
Multiplication	$Area = Average\ step \cdot amount = \pi r \cdot r = \pi r^2$	
Integration	$Area = \int actual\ steps = \int 2\pi r = \pi r^2$	

Division spits back the averaged-sized ring in our pattern. The derivative gives a formula ($2\pi r$) that describes *every* ring (just plug in r). Similarly, multiplication lets us scale up the average element (once we've found it) into the full amount. Integrals let us add up the pattern directly.

Sometimes we want to use the average item, not the fancy calculus steps, because it's a simpler representation of the whole ("What's the average transaction size? I don't need the full list."). That's fine, as long as it's a conscious choice.

6.2 Better Formulas

If calculus provides better, more-specific version of multiplication and division, shouldn't we rewrite formulas with it? You bet.

Algebra	Calculus
$distance = speed \cdot time$	$distance = \int speed \, dt$
$speed = \dfrac{distance}{time}$	$speed = \dfrac{d}{dt} distance$
$area = height \cdot width$	$area = \int height \, dw$
$weight = density \cdot length \cdot width \cdot height$	$weight = \iiint density \, dx \, dy \, dz$

An equation like *distance = speed·time* explains how to find total distance assuming an average speed. An equation like $distance = \int speed \, dt$ tells us how to find total distance by breaking time into instants (split along the "t" axis), and accumulating the (potentially unique) distance traveled each instant (*speed·dt*).

Similarly, $speed = \frac{d}{dt} distance$ explains that we can split our trajectory into time segments, and the (potentially unique) amount we moved in that time slice was the speed.

The overused "integrals are area under the curve" explanation becomes more clear. Multiplication, because it deals with static quantities, can only measure the area of rectangles. Integrals let measurements curve and undulate as we go: we'll add their contribution, regardless.

A series of multiplications becomes a series of integrals (called a triple integral). It's beyond this primer, but your suspicion was correct: we can mimic the multiplications and integrate several times in a row.

Math, and specifically calculus, is the language of science because it describes relationships extremely well. When I see a formula with an integral or derivative, I mentally convert it to multiplication or division (remembering we can handle differently-sized elements).

6.3 Better Algebra

Algebra lets us start with one fact and systematically work out others. Imagine I want to know the area of an unknown square. I can't measure the area, but I overheard someone saying it was 13.3 inches on a side.

Algebra	Thinking Process
Area of square = ?	The area of this square is unknown...
$\sqrt{Area} = 13.3$	...but I know the square root.
$\left(\sqrt{Area}\right)^2 = (13.3)^2$	Square both sides...
Area = 176.89	...and I can recreate the original area

Remember learning that along with add/subtract/multiply/divide, we could take powers and roots? We added two new ways to transform an equation.

Well, calculus extends algebra with two more operations: integrals and derivatives. Now we can work out the area of a circle, algebra-style:

Algebra + Calculus	Thinking Process
Area of circle =?	The area of a circle is unknown...
$\frac{d}{dr}Area = 2\pi r$	...but I know it splits into rings (along the radius)
$\int \frac{d}{dr}Area = \int 2\pi r$	Integrate both sides...
$Area = \pi r^2$	...and I can recreate the original area

The abbreviated notation helps see the big picture. If the integrand only uses a single variable (as in $2\pi r$), we can assume we're using dr from $r = 0$ to $r = r$. This helps us think of integrals and derivatives like squares and square roots: operations that cancel!

It's pretty neat: "gluing together" and "splitting apart" should behave like opposites, right?

With our simpler notation, we can write $\int \frac{d}{dr}Area = Area$ instead of the bulky $\int_0^r \left(\frac{d}{dr}Area\right) dr = Area$.

6.4 Learning The Rules

With arithmetic, we learned special techniques for combining whole numbers, decimals, fractions, and roots/powers. Even though $3 + 9 = 12$, we can't assume $\sqrt{3} + \sqrt{9} = \sqrt{12}$.

Similarly, we need to learn the rules for how integrals/derivatives work when added, multiplied, and so on. Yes, there are fancy rules for special categories (what to do with e^x, natural log, sine, cosine, etc.), but I'm not concerned with that. Let's get extremely comfortable with the basics. The fancy stuff can wait.

Chapter 7

Seeing How Lines Work

Let's start by analyzing a fairly simple pattern, a line:

$$f(x) = 4x$$

In everyday terms, we enter an input, x, and get an output, $f(x)$. Suppose we're buying fencing. For every foot we ask for (the input, x), it costs us $4 (the output, $f(x)$). 3 feet of fence would cost $12. Fair enough.

Notice the abstract formula $f(x) = 4x$ only considers numerical quantities, but not their units (feet, dollars, etc.). We could write that a foot of fencing costs 400 pennies ($f(x) = 400x$) and it's up to *us* to realize it's the same scenario. A big gotcha in Calculus is realizing x, dx and friends have sizes – but not units – which we eventually interpret as area, volume, dollars, pennies, etc. In Math Land, everything is a number.

7.1 Finding the Derivative Of A Line

The derivative of a pattern, $\frac{d}{dx} f(x)$, is the sequence of slices we get as we change an input variable (x is the natural choice here). How do we figure out the sequence of steps?

Well, I imagine going to Home Depot and pestering the clerk:

> You: I'd like some lumber please. What will it run me?
>
> Clerk: How much do you want?
>
> You: Um. . . 1 foot, I think.
>
> Clerk: That'll be $4. Anything else I can help you with?
>
> You: Actually, it might be 2 feet.
>
> Clerk: That'll be $8. Anything else I can help you with?
>
> You: It might be 3 feet.
>
> Clerk: *(sigh)* That'll be $12. Anything else I can help you with?

You: How about 4 feet?

We have a relationship ($f(x) = 4x$) and investigate it by changing the input a tiny bit. We see if there's a change in output (there is!), then we change the input again, and so on.

In this case, it's clear that an additional foot of fencing raises the cost by $4. So we've just determined the derivative to be a constant 4, right?

Not so fast. Sure, we thought about the process and worked it out, but let's be a little more organized (not every pattern is so simple). Can we describe our steps?

1. Get the current output, $f(x)$. In our case, $f(1) = 4$.

2. Step forward by dx (1 foot, for example)

3. Find the new amount, $f(x+dx)$. In our case, it's $f(1+1) = f(2) = 8$.

4. Compute the difference: $f(x+dx) - f(x)$, or 8 - 4 = 4

Ah! The difference between the next step and the current one is the size of our slice. For $f(x) = 4x$ we have:

$$f(x+dx) - f(x) = 4(x+dx) - 4(x) = 4 \cdot dx$$

Increasing length by dx increases the cost by $4 \cdot dx$.

That statement is true, but a little awkward: it talks about the total change. Wouldn't it be better to have a ratio, such as "cost per foot"?

We can extract the ratio with a few shortcuts:

- dx = change in our input

- df = resulting change in our output, $f(x+dx) - f(x)$

- $\frac{df}{dx}$ = ratio of output change to input change

In our case, we have

$$\frac{df}{dx} = \frac{4 \cdot dx}{dx} = 4$$

Notice how we express the derivative as $\frac{df}{dx}$ instead of $\frac{d}{dx} f(x)$. What gives? It turns out there's a few different versions we can use.

Think about the various ways we express multiplication:

- Times symbol: 3 × 4 (used in elementary school)

- Dot: 3 · 4 (used in middle school)

- Implied multiplication with parentheses: $(x+4)(x+3)$

- Implied multiplication with a space: $2\pi r \, dr$

The more subtle the symbol, the more we focus on the relationship between the quantities; the more visible the symbol, the more we focus on the computation.

The notation for derivatives is similar:

Derivative Symbol	Mindset
$f'(x) \quad \dot{f} \quad \frac{d}{dx}f$	Think about the resulting step-by-step pattern
$\frac{df}{dx}$	Think about the actual ratio of changes

Some versions, like $f'(x)$, remind us the sequence of steps is a variation of the original pattern. Notation like $\frac{df}{dx}$ puts us into detail-oriented mode, thinking about the ratio of output change relative to the input change ("What's the cost per additional foot?").

Remember, the derivative is a complete description of all the steps, but it can be evaluated at a certain point to find the step there: What is the additional cost/foot when $x = 3$? In our case, the answer is 4.

Here's what the computer returns for this problem:

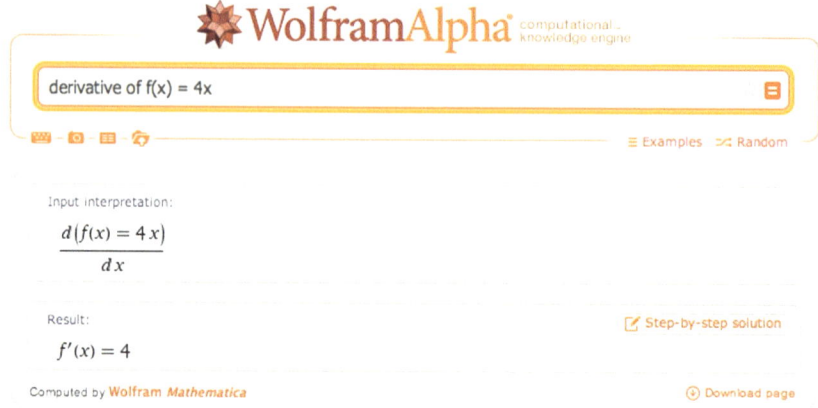

Nice! As we suspected, the pattern $f(x) = 4x$ changes by a constant 4 as we increase x.

7.2 Finding The Integral Of A Constant

Now let's work in the other direction: given the sequence of steps, can we find the size of the original pattern?

In our fence-building scenario, it's fairly straightforward. Solving

$$\frac{df}{dx} = 4$$

means answering "What pattern has an output change of 4 times the input change?".

Well, we've just seen that $f(x) = 4x$ results in $f'(x) = 4$. So, if we're given $f'(x) = 4$, we can guess the original function must have been $f(x) = 4x$.

I'm pretty sure we're right (what else could the integral of 4 be?), but let's compare this with the computer:

WolframAlpha computational knowledge engine

integral of 4 from x = 0 to x = x

Definite integral:
$$\int_0^x 4\,dx = 4x$$

Indefinite integral:
$$\int 4\,dx = 4x + \text{constant}$$

Computed by Wolfram Mathematica

Whoa – there's two different answers (definite and indefinite). Why? Well, there's many functions that could increase cost by $4/foot! Here's a few:

- Cost = $4 per foot, or $f(x) = 4x$
- Cost = $4 + $4 per foot, or $f(x) = 4 + 4x$
- Cost = $10 + $4 per foot, or $f(x) = 10 + 4x$

There could be a fixed per order fee, with the fence cost added in. All the equation $f'(x) = 4$ says is that each *additional* foot of fencing is $4, but we don't know the starting conditions.

- The **definite integral** tracks the accumulation of a set amount of slices. The range can be numbers, such as $\int_0^{13} 4$, which measures the slices from x=0 to x=13 ($13 \cdot 4 = 52$). If the range includes a variable (0 to x), then the accumulation will be an equation ($4x$).

- The **indefinite integral** finds the actual formula that created the pattern of steps, not just the accumulation in that range. It's written with just an integral sign: $\int f(x)$. And as we've seen, the possibilities for the original function should allow for a starting offset of C.

The notation for integrals can be fast-and-loose, and it's confusing. Are we looking for an accumulation, or the original function? Are we leaving out dx? These details are often omitted, so it's important to feel what's happening.

7.3 The Secret: We Can Work Backwards

The little secret of integrals is that we don't need to solve them directly. Instead of trying to glue slices together to find out their area, we just learn to *recognize* the derivatives of functions we've already seen.

If we know the derivative of 4x is 4, then if someone asks for the *integral* of 4, we can respond with "4x" (plus C, of course). It's like memorizing the squares of numbers, not the square roots. When someone asks for the square root of 121, dig through and remember that 11 × 11 = 121.

An analogy: Imagine an antiques dealer who knows the original vase just from seeing a pile of shards.

How does he do it? Well, he takes replicas in the back room, drops them, and looks at the pattern of pieces. Then he comes to your pile and says "Oh, I think this must be a Ming Dynasty Vase from the 3rd Emperor."

He doesn't try to glue your pile back together – he's just seen that exact vase break before, and your pile looks the same!

Now, there may be piles he's never seen, that are difficult or impossible to recognize. In that case, the best he can do is to just add up the pieces (with a computer, most likely). He might determine the original vase weighed 13.78 pounds. That's a data point, fine, but it's not as nice as knowing *what* the vase was before it shattered.

This insight was never really explained to me: it's painful to add up (possibly changing) steps directly, especially when the pattern gets complicated. So, just learn to recognize the pattern from the derivatives we've already seen.

7.4 Getting To Better Multiplication

Gluing together equally-sized steps looks like regular multiplication, right? You bet. If we wanted 3 steps (0 to 1, 1 to 2, 2 to 3) of size 2, we might write:

$$\int_0^3 2 \, dx = 6$$

Again, this is a fancy way of saying "Accumulate 3 steps of size 2: what do you get in total?". We are time-lapsing a sequence of equal changes.

Now, suppose someone asks you to add 2 + 2 + 2 + 2 + 2 + 2 + 2 + 2 + 2 + 2 + 2 + 2 + 2. You might say: *Geez, can't you write it more simply? You know, something like:*

$$\int_0^{13} 2 \, dx = 26$$

7.5 Creating The Abstract Rules

Have an idea how linear functions behave? Great. We can make a few abstract rules – like working out the rules of algebra for ourselves.

If we know our output is a scaled version of our input ($f(x) = ax$), the derivative (pattern of changes) is

$$\frac{d}{dx} a \cdot x = a$$

and the integral (pattern of accumulation) is

$$\int a = ax + C$$

That is, the ratio of each output step to each input step is a constant a (4, in our examples above). And now that we've broken the vase, we can work backwards: if we accumulate steps of size a, they must have come from a pattern similar to $a \cdot x$ (plus C, of course).

Notice how I wrote $\int a$ and not $\int a\, dx$ – I wanted to focus on a, and not details like the width of the step (dx). Part of calculus is learning to expose the right amount of detail.

One last note: if our output does not react *at all* to our input (we'll charge you a constant $2 no matter how much you buy... including nothing!) then "steps" are a constant 0:

$$\frac{d}{dx} a = 0$$

In other words, there is no difference in the before-and-after measurement. Now, a pattern may have an *occasional* zero slice, if it stands still for a moment. That's fine. But if *every* slice were zero, it means our pattern never changes.

There are a few subtleties down the road, but let's learn to say "Me want food" before "Verily, I hunger".

CHAPTER 8

PLAYING WITH SQUARES

We've seen how lines behave: they change the same amount with each step. Now let's try a more complex function like $f(x) = x^2$. It's a more detailed scenario, so let's visualize it.

Imagine you're building a square garden, to plant veggies and enjoy cucumbers in a few months. You're not sure how large to make it. Too small, and there's not enough food, but too large, and you'll draw the attention of the veggie mafia.

Your plan is to build the garden incrementally, foot-by-foot, until it looks right. Let's say you start from scratch and build up to a 10 × 10 plot:

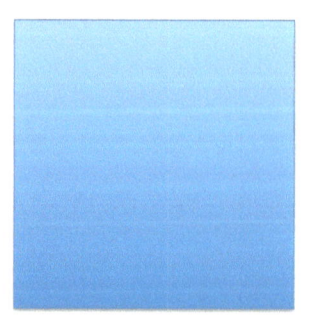

To the untrained eye, you have single a 10 × 10 garden, which uses 40 feet of perimeter fencing (10 × 4) and 100 square feet of topsoil (10 × 10). (Assume topsoil is sold by the square foot, with a standard thickness.)

8.1 Bring On The Calculus

That's it? The analysis just figures out the current perimeter and square footage? No way.

By now, you should be clamoring to use X-Ray and Time-Lapse vision to see what's happening under the hood. Why settle for a static description when we can know the step-by-step description too?

We can analyze the behavior of the perimeter pretty easily:

$$Perimeter = 4x$$

$$\frac{d}{dx}\text{Perimeter} = 4$$

The change in perimeter ($\frac{dP}{dx}$) is a constant 4. For every 1-foot increase in x, we have a 4-foot jump in the perimeter.

We can visualize this process. As the square grows, we push out the existing sides and add 4 corner pieces (in yellow):

The visual is helpful, but not required. After our exposure to how lines behave, we can glance at an equation like $p = 4x$ and realize that p jumps by 4 whenever x jumps by 1.

8.2 Changing Area

Now, how does area change? Since squares are fairly new, let's X-Ray the shape as it grows:

We can write out the size of each jump, like so:

CHAPTER 8. PLAYING WITH SQUARES

x	x^2	Jump to next square
0	0	1 ($1^2 - 0^2 = 1$)
1	1	3 ($2^2 - 1^2 = 3$)
2	4	5 ($3^2 - 2^2 = 5$)
3	9	7
4	16	9
5	25	11
6	36	13
7	49	15

Now that's interesting. The gap from 0^2 to 1^2 is 1. The gap from 1^2 to 2^2 is 3. The gap from 2^2 to 3^2 is 5. And so on – the odd numbers are sandwiched between the squares! What's going on?

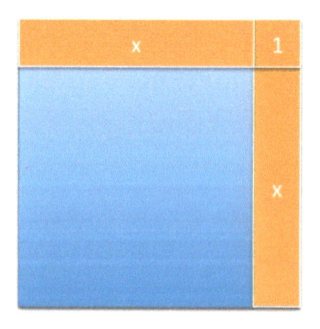

Ah! Growing to the next-sized square means we've added a horizontal and vertical strip ($x + x$) and a corner piece (1). If we currently have a square with side x, the jump to the next square is $2x + 1$. (If we have a 5 × 5 square, getting to a 6 × 6 will be a jump of $2(5) + 1 = 11$. And yep, $36 - 25 = 11$.)

Again, the visualization was nice, but it took effort. Algebra can simplify the process.

In this setup, if we set our change to $dx = 1$, we get

$$\begin{aligned} df &= f(x+1) - f(x) \\ &= (x+1)^2 - x^2 \\ &= (x^2 + 2x + 1) - x^2 \\ &= 2x + 1 \end{aligned}$$

Algebra predicts the size of the slices without a hitch.

8.3 Integrals and the Veggie Mafia

The derivative takes a shape, a direction to cut, and finds a pattern of slices. Can we work backwards, from the slices to the shape? Let's see.

Suppose the veggie mafia spies on your topsoil and fencing orders. They can't see your garden directly, but what can they deduce from your purchases?

Let's say they observe a constant amount of fencing being delivered (4, 4, 4, 4...) but *increasing* orders of topsoil (1, 3, 5, 7, 9, 11...). What can they work out?

A low-level goon might just add up the total amount accumulated (the definite integral): "Heya boss, looks like they've built some garden with a total perimeter of 40-feet, and total area of 100 square feet."

But that's not good enough! The goon doesn't know the shape you're trying to build. He saw order after order go by without noticing the deeper pattern.

The crime boss is different: he wants the *indefinite* integral, the pattern you are following. He's savvy enough to track the pattern as the orders come in: "The area is increasing 1, 3, 5, 7... that's following a $2x+1$ area increase pattern!"

Now, there are likely many shapes that could grow their area by $2x+1$. But, combined with a constant perimeter increase of 4, he suspects you're making a square garden after a few deliveries.

How does the godfather do it? Again, by working backwards. He's split apart enough shapes (triangles, squares, rectangles, etc.) that he has a large table of before-and-afters, just like the antiques dealer.

When he sees a change of $2x+1$, a square (x^2) is a strong candidate. Another option might be a right triangle with sides x and $2x$. Its area equation is $\frac{1}{2}x \cdot 2x = x^2$, so the area would change the same as a square.

And when he sees a perimeter change of a steady 4, he knows the perimeter must be $4x$. Ah! There aren't too many shapes with both properties: a square is his guess. (With enough practice, you start to recognize common patterns; tools like Wolfram Alpha can help.)

Now suppose your orders change: your fencing deliveries drop to (2, 2, 2, 2...) and your topsoil orders change to (20, 20, 20, 20). What's going on? Make a guess if you like.

Ready?

The veggie boss figures you've moved to a *rectangular* garden, with one side determined by x, and the other side a fixed 20 feet, for a 20-by-x rectangle.

Does this guess work? Assuming this is the pattern, let's measure the perimeter, area, and how they change:

$$\begin{aligned} \text{Perimeter} &= 20 + 20 + x + x = 40 + 2x \\ \frac{d}{dx}\text{Perimeter} &= 2 \\ \text{Area} &= 20x \\ \frac{d}{dx}\text{Area} &= 20 \end{aligned}$$

Wow, it checks out: the changes in perimeter and area match the patterns (2, 2, 2...) and (20, 20, 20...). No wonder he's the godfather.

Lastly, what if the godfather saw topsoil orders of (5, 7, 9, 11, 13)? He might assume you're still building a square ($2x+1$ pattern), but you *started* with a 2×2 garden. Your first area jump was by 5, which would have happened if x was already 2 (solve $2x+1=5$ and we see $x=2$).

The mob boss is a master antiques dealer: he sees the pattern in the pieces you're bringing and quickly determines the original shape (indefinite integral). The henchman can only tell you the running totals so far (definite integral).

8.4 Wrapping It All Up

It looks like we're ready for another rule, to explain how squares change. If we leave dx as it is, we can write:

$$\begin{aligned} \frac{d}{dx}x^2 &= \frac{f(x+dx)-f(x)}{dx} \\ &= \frac{(x+dx)^2-(x)^2}{dx} \\ &= \frac{x^2+2x \cdot dx+(dx)^2-x^2}{dx} \\ &= \frac{2x \cdot dx+(dx)^2}{dx} \\ &= 2x+dx \end{aligned}$$

Ok! That's the abbreviated way of saying "Grow by two sides and the corner". Let's plug this into the computer to check:

WolframAlpha computational knowledge engine

derivative of x^2

Examples   Random

Derivative: Step-by-step solution

$$\frac{d}{dx}(x^2) = 2x$$

Uh oh! We hand-computed the derivative of x^2 as $2x+dx$ (which is usually $2x+1$), but the computer says it's just $2x$.

But isn't the difference from 4^2 to 5^2 exactly $25-16=9$, and not 8? What happened to that corner piece? *The mystery continues.*

CHAPTER 9

WORKING WITH INFINITY

Last time, we manually worked on the derivative of x^2 as $2x + 1$. But the official derivative, according to the calculator, was $2x$. What gives?

The answer relies on the concept of infinite accuracy. Infinity is a fascinating and scary concept – there are entire classes (Analysis) that study it. We'll avoid the theoretical nuances: our goal is a practical understanding of how infinity helps us with Calculus.

9.1 Insight: Sometimes Infinity Can Be Measured

Here's a quick brainteaser for you. Two friends are 10 miles apart, moving towards each other at 5mph each. A mosquito files quickly between them, touching one person, then the other, on and on, until the friends high-five and the mosquito is squished.

Let's say the mosquito travels a zippy 20mph as it goes. Can you figure out how far it flew before its demise?

Yikes. This one is tricky: once the mosquito leaves the first person, touches the second, and turns around... the first person has moved closer! We have an infinite number of ever-diminishing distances to add up. The question seems painfully difficult to solve, right?

Well, how about this reasoning: from the perspective of the people walking, they're going to walk for an hour total. After all, they start 10 miles apart, and the gap shrinks at 10 miles per hour (5mph + 5mph). Therefore, the mosquito must be flying for an hour, and go 20 miles.

Whoa! Did we just find the outcome of a process with an infinite number of steps? I think so!

9.2 Splitting A Whole Into Infinite Parts

It's time to turn our step-by-step thinking into overdrive. Can we think about a finite shape being split into infinite parts?

- In the beginning of the course, we saw a circle could be split into rings. How many? Well, an infinite number!

- A number line can be split into an *infinite* number of neighboring points. How many decimals would you say there are between 1.0 and 2.0?

- The path of a mosquito can be seen as a whole, or a journey subdivided into an infinite number of segments.

When we have two viewpoints (the mosquito, and the walkers), we can pick the one that's easier to work with. In this case, the walker's holistic viewpoint is simpler. With the circle, it's easier to think about the rings themselves. It's nice to have both options available.

Here's another example: can you divide a cake into 3 equal portions, by only cutting into quarters?

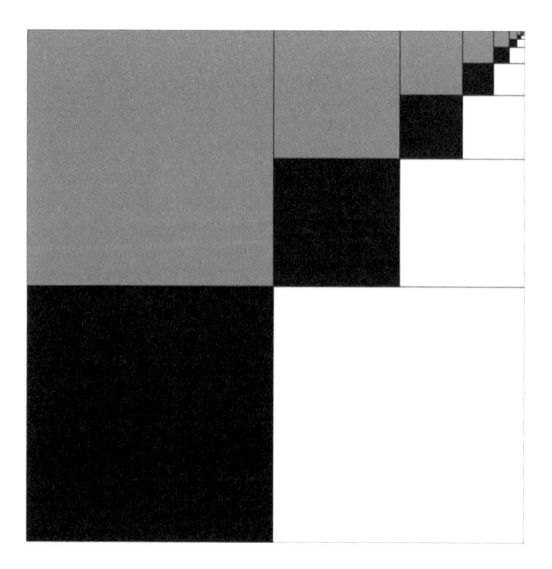

It's a weird question... but possible! Cut the entire cake into quarters. Share 3 pieces and leave 1. Cut the remaining piece into quarters. Share 3 pieces, leave 1. Keep repeating this process: at every step, everyone has received an equal share, and the remaining cake will be split evenly as well. Wouldn't this plan maintain an even split among 3 people?

We're seeing the intuition behind infinite X-Ray and Time-lapse vision: zooming in to turn a whole into an infinite sequence. At first, we might think dividing something into infinite parts requires each part to be nothing. But, that's not right: the number line can be subdivided infinitely, yet there's a finite gap between 1.0 and 2.0.

9.3 Two Fingers Pointing At The Same Moon

Why can we understand variations of the letter A, even when pixelated?

a a a

Even though the rendering is different, we see the **idea being pointed to**. All three versions, from perfectly smooth to jagged, create the same letter A in our heads (or, are you unable to read words when written out with rectangular pixels?). An infinite sequence can point to the same result we'd find if we took it all at once.

In calculus, there are detailed rules about how to find what result an infinite set of steps points to. And, there are certain sequences that cannot be worked out. But, for this primer, we'll deal with functions that behave nicely.

We're used to jumping between *finite* representations of the same idea (5 − V = IIII). Now we're seeing we can convert between a finite and *infinite* representation of an idea, similar to $\frac{1}{3} = .333\ldots = .3 + .03 + .003 + \ldots$.

When we turned a circle into a ring-triangle, we said "The infinitely-many rings in our circle can be turned into the infinitely-many boards that make up a triangle. And the resulting triangle is easy to measure."

Unroll the Rings

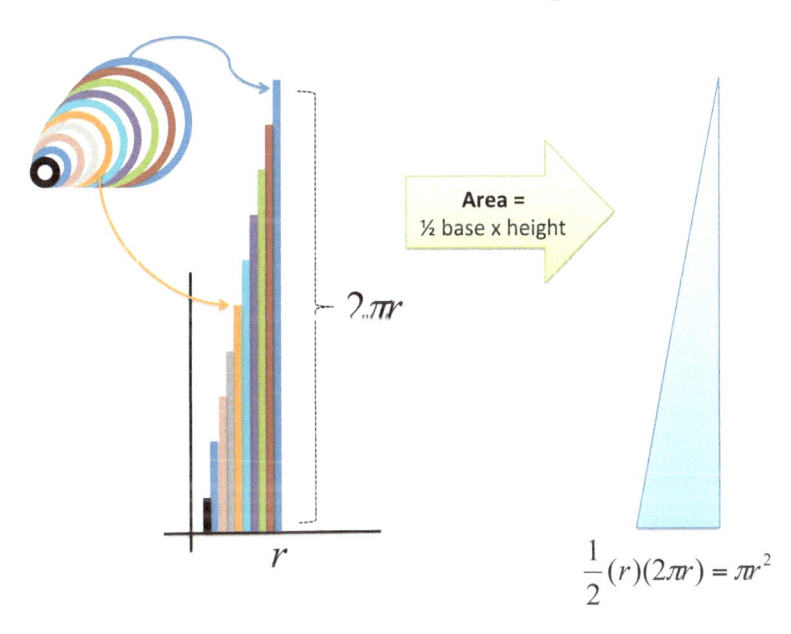

Area = ½ base × height

$$\frac{1}{2}(r)(2\pi r) = \pi r^2$$

Today's goal isn't to become an expert on infinity. It's to intuitively appreciate a few practical conclusions:

- Infinitely many parts can combined to a finite result, if they decrease fast enough

CHAPTER 9. WORKING WITH INFINITY

- A process with limited (but improving) precision can point to the same result as one with infinite precision

In Calculus terms, this means the conclusions drawn from our finite (but growing) sequence of steps can be trusted[1].

---

[1] Calculus is a powerful but not flawless tool. Jumpy, artificial patterns trip it up and can't be analyzed. Luckily, most naturally-occurring patterns can be.

CHAPTER 10

THE THEORY OF DERIVATIVES

The last lesson showed that an infinite sequence of steps could have a finite conclusion. Let's put it into practice, and see how breaking change into infinitely small parts can point to the true amount.

10.1 Analogy: Measuring Heart Rates

Imagine you're a doctor trying to measure a patient's heart rate while exercising. You put a guy on a treadmill, strap on the electrodes, and get him running. The machine spits out 180 beats per minute. That must be his heart rate, right?

Nope. That's his heart rate *when observed by doctors and covered in electrodes*. Wouldn't that scenario be stressful? And what if your Nixon-era electrodes get tangled on themselves, and tug on his legs while running?

Ah. We need the electrodes to get *some* measurement. But, right afterwards, we need to remove the effect of the electrodes themselves. For example, if we measure 180 bpm, and knew the electrodes added 5 bpm of stress, we'd know the true heart rate was 175.

The key is making the knowingly-flawed measurement, getting a reading, then correcting it as if the instrument was never there.

10.2 Measuring the Derivative

Measuring the derivative is just like putting electrodes on a function and making it run. For $f(x) = x^2$, we stick an electrode of +1 onto it, to see how it reacted:

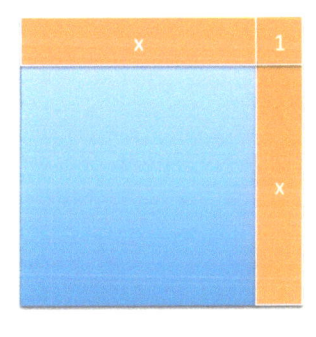

The horizontal stripe is the result of our change applied along the top of the shape. The vertical stripe is our change moving along the side. And what's the corner?

It's part of the horizontal change interacting with the vertical one! This is an electrode getting tangled in its own wires, a measurement artifact that needs to go.

10.3 Throwing Away Artificial Results

The founders of calculus intuitively recognized which components of change were "artificial" and just threw them away. They saw that the corner piece was the result of our test measurement interacting with itself, and shouldn't be included.

In modern times, we created official theories about how this is done:

- Limits: We let the measurement artifacts get smaller and smaller until they effectively disappear (cannot be distinguished from zero).

- Infinitesimals: Create a new type of number that lets us try infinitely-small change on a separate, tiny number system. When we bring the result back to our regular number system, the artificial elements are removed.

The are entire classes that explore these theories. The practical upshot is realizing *how* to take a measurement and then throwing away the parts we don't need.

Here's how the derivative is defined using limits:

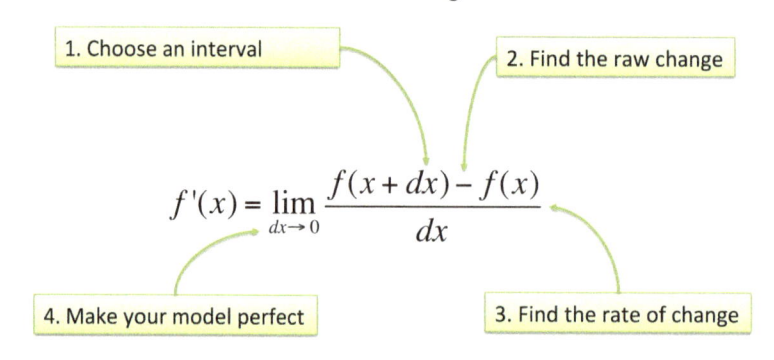

1. Choose an interval
2. Find the raw change

$$f'(x) = \lim_{dx \to 0} \frac{f(x+dx) - f(x)}{dx}$$

4. Make your model perfect
3. Find the rate of change

Step	Example
Start with function to study	$f(x) = x^2$
1. Increase the input by dx, a sample change	$f(x+dx) = (x+dx)^2 = x^2 + 2x \cdot dx + (dx)^2$
2. Find the resulting increase in output, df	$df = f(x+dx) - f(x) = 2x \cdot dx + (dx)^2$
3. Find the ratio of output change to input change	$\frac{df}{dx} = \frac{2x \cdot dx + (dx)^2}{dx} = 2x + dx$
4. Throw away any measurement artifacts	$2x + dx \overset{dx=0}{\Longrightarrow} 2x$

Wow! We found the official derivative for $\frac{d}{dx} x^2$ on our own:

Now, a few questions:

- **Why do we measure $\frac{df}{dx}$, and not the actual change df?** Think of df as the entire change that happened as we took a step. For easy comparison to other functions, we typically want the "per step" change $\frac{df}{dx}$. (This is like comparing jobs by dollars/hour instead of by salary, or cars by miles-per-gallon instead of gallons used.) Sometimes the total change is helpful to consider, and we can rewrite $\frac{df}{dx} = 2x$ as $df = 2x \cdot dx$.

- **How can we just set dx to zero at the end?** I see dx as the size of the instrument used to measure the change in a function. After we have the measurement with a real instrument ($\frac{df}{dx} = 2x + dx$), we figure out what the measurement would be if the instrument were perfect and did not interfere ($\frac{df}{dx} = 2x + 0 = 2x$).

- **But isn't the $2x+1$ pattern correct?** The whole numbers (integers) are separated by an interval of 1, so assuming $dx = 1$ (and not letting it disappear) is accurate: $2x+1$ correctly predicts the gap of 5 between 2^2 and 3^2. However, decimals (real numbers) don't have a fixed interval between neighbors. $2x$ is the ideal estimate for the rate of change between 2^2 and the infinitely-close number that follows – not 2.0001, or 2.0000000001, but whatever unnamed number comes next. Said another way, if dx doesn't disappear, we're saying the real numbers have a fixed interval between them, like the integers.

- **If there's no '+1', when is the corner filled in?** Think about the change in area, and not the specifics of the diagram. The corner overestimates how much growth happens on *this step* (i.e., the radar clocked us at $2x+1$ but we're only growing by $2x$). But we're still moving and make progress.

  I imagine a square that grows by bulging out its sides ($x + x = 2x$), then absorbing the new area to make a larger square. The new size is larger, but not *quite* big enough to fill in the corner exactly. It's ok, because this process will still encompass the necessary area over time. $2x+1$ overestimates our growth because it assumes the horizontal and vertical slices interact to create the corner piece.

Practical conclusion: We can start with a knowingly-flawed measurement, $f'(x) \sim 2x + dx$, and deduce the perfect result it points to: $f'(x) = 2x$. The theories of exactly *how* we throw away dx aren't necessary to master today. The key is realizing there are measurement artifacts – the shadow of the camera in the photo – that must be removed to accurately describe the true behavior.

(Still shaky about exactly how dx can appear and disappear? You're in good company. This question took top mathematicians decades to resolve. Here's a deeper discussion of how the theory works.)

CHAPTER 11

THE FUNDAMENTAL THEOREM OF CALCULUS (FTOC)

The Fundamental Theorem of Calculus is the big aha! moment, and something you might have noticed all along:

- X-Ray and Time-Lapse vision let us see an existing pattern as an accumulated sequence of changes
- The two viewpoints are opposites: X-Rays break things apart, Time-Lapses put them together

This might seem "obvious", but it's only because we've explored several examples. Is it truly obvious that we can separate a circle into rings to find the area? The Fundamental Theorem of Calculus gently reminds us we have a few ways to look at a pattern. (*"Might I suggest the ring-by-ring viewpoint? Makes things easier to measure, I think."*)

11.1 Part 1: Shortcuts For Definite Integrals

If derivatives and integrals are opposites, we can sidestep the laborious accumulation process found in definite integrals.

For example, what is $1 + 3 + 5 + 7 + 9$? The hard way, computing the definite integral directly, is to add up the items directly. (*What about 50 items? 500?*)

The easy way is to realize this pattern of numbers comes from a growing square. We know the last change (+9) happens at $x = 4$, so we've built up to a 5×5 square. Therefore, the sum of the entire sequence is 25:

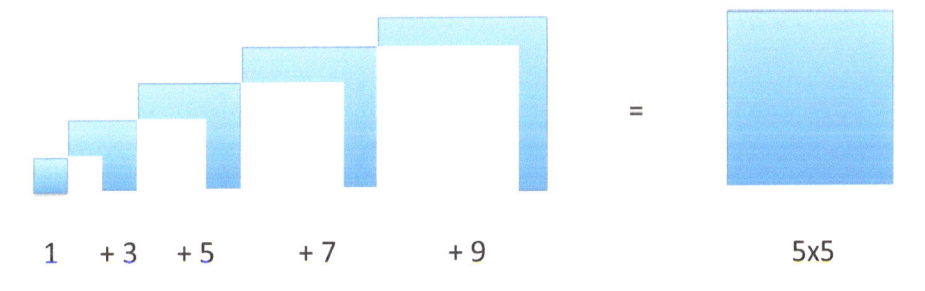

Neat! If we have the original pattern, we have a shortcut to measure the size of the steps.

How about a partial sequence like 5 + 7 + 9? Well, just take the total accumulation and subtract the part we're missing (in this case, the missing 1 + 3 represents a missing 2×2 square).

| 5 | + 7 | + 9 | | 5x5 | - | 2x2 |

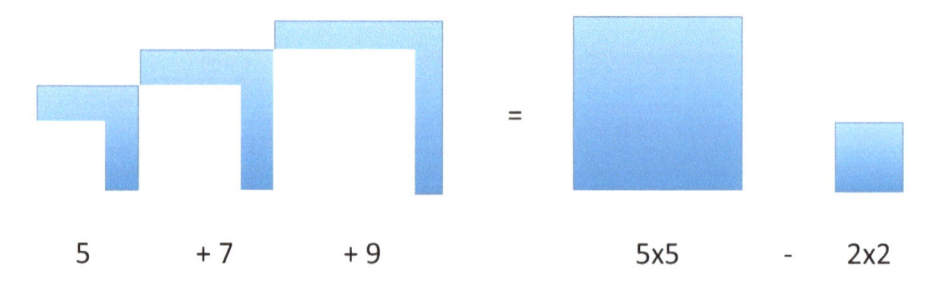

And yep, the sum of the partial sequence is: 5×5 - 2×2 = 25 - 4 = 21.

I hope the strategy clicks for you: avoid manually computing the definite integral by finding the original pattern.

Here's the first part of the FTOC in fancy language. If we have pattern of steps and the original pattern, the shortcut for the definite integral is:

$$\int_a^b steps(x)dx = Original(b) - Original(a)$$

Intuitively, I read this as "Adding up all the changes from a to b is the same as getting the difference between a and b". Formally, you'll see $f(x) = steps(x)$ and $F(x) = Original(x)$, which I think is confusing. Label the steps as steps, and the original as the original.

Why is this cool? The definite integral is a gritty mechanical computation, and the indefinite integral is a nice, clean formula. Just take the difference between the endpoints to know the net result of what happened in the middle! (That makes sense, right?)

11.2 Part 2: Finding The Indefinite Integral

Ok. Part 1 said that if we have the original function, we can skip the manual computation of the steps. But how do we find the original?

FTOC Part Deux to the rescue!

Let's pretend there's *some* original function (currently unknown) that tracks the accumulation:

$$Accumulation(x) = \int_a^b steps(x)dx$$

The FTOC says the *derivative* of that magic function will be the steps we have:

$$Accumulation'(x) = steps(x)$$

Now we can work backwards. If we can find some random function, take its derivative, notice that it matches the steps we have, we can use that function as our original!

Skip the painful process of thinking about *what* function could make the steps we have. Just take a bunch of them, break them, and see which matches up. It's our vase analogy, remember? The FTOC gives us "official permission" to work backwards. In my head, I think "The next step in the total accumulation is our current amount! That's why the derivative of the accumulation matches the steps we have."

Technically, a function whose derivative is equal to the current steps is called an anti-derivative (One anti-derivative of 2 is $2x$; another is $2x + 10$). The FTOC tells us any anti-derivative will be the original pattern (+C of course).

This is surprising – it's like saying everyone who *behaves* like Steve Jobs *is* Steve Jobs. But in Calculus, if a function splits into pieces that match the pieces we have, it was their source.

The practical conclusion is **integration and differentiation are opposites**. Have a pattern of steps? Integrate to get the original. Have the original? Differentiate to get the pattern of steps. Jump back and forth as many times as you like.

11.3 Next Steps

Phew! These lessons were theory-heavy, to give an intuitive foundation for topics in an Official Calculus Class. The key insights are:

- **Infinity:** A finite result can be viewed with a sequence of infinite steps.

- **Derivatives:** We can take a knowingly-flawed measurement and find the ideal result it refers to.

- **Fundamental Theorem Of Calculus:** The original function lets us skip adding up a gajillion small pieces.

In the upcoming lessons, we'll work through a few famous calculus rules and applications. The real goal will be to figure out, for ourselves, how to make this happen:

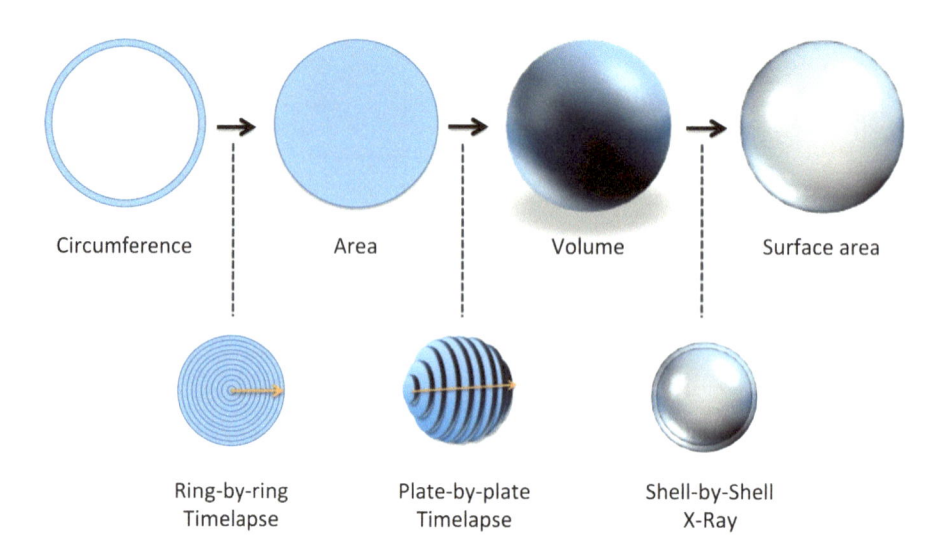

CHAPTER 11. THE FUNDAMENTAL THEOREM OF CALCULUS (FTOC)

Circumference	Area	Volume	Surface area
	Ring-by-ring Timelapse	Plate-by-plate Timelapse	Shell-by-Shell X-Ray

By now, we have an idea that the strategy above is possible. By the last chapter, you'll be able to walk through the exact calculations on your own.

CHAPTER 12

THE BASIC ARITHMETIC OF CALCULUS

Remember learning arithmetic? After seeing how to multiply small numbers, we learned how to multiply numbers with several digits:

$$13 \times 15 = (10+3)(10+5) = 100 + 30 + 50 + 15$$

We can't just combine the first digits (10×10) and the second (3×5) and call it done. We have to walk through the cross-multiplication.

Calculus is similar. If we have the whole function, we can blithely say that $f(x)$ has derivative $f'(x)$. But that isn't illuminating, or explaining what happens behind the scenes.

If we can describe our function in terms of a building block x (such as $f(x) = 3x^2 + x$), then we should be able to find the derivative, the pattern of changes, in terms of that same building block. If we have two types of building blocks ($f = a \cdot b$), we'll get the derivative in terms of those two building blocks.

Here's the general strategy:

- Imagine a scenario with a few building blocks ($area = length \cdot width$)

- Let every component change

- Measure the change in the overall system

- Remove the measurement artifacts (our instruments interfering with each other)

Once we know how systems break apart, we can reverse-engineer them into the integral (yay for the FTOC!).

12.1 Addition

Let's start off easy: how does a system with two added components behave?

In the real world, this could be sending two friends (Frank and George) to build a fence. Let's say Frank gets the wood, and George gets the paint. What's the total cost?

$$Total = Frank's\ cost + George's\ cost$$

CHAPTER 12. THE BASIC ARITHMETIC OF CALCULUS

$$t(x) = f(x) + g(x)$$

The derivative of the entire system, $\frac{dt}{dx}$, is the cost per additional foot. Intuitively, we suspect the total increase is the sum of the increases in the parts:

$$\frac{dt}{dx} = \frac{df}{dx} + \frac{dg}{dx}$$

That relationship makes sense, right? Let's say Frank's cost is $3/foot for the wood, and George adds $0.50/foot for the paint. If we ask for another foot, the total cost will increase by $3.50.

Here's the math for that result:

- Original: $f + g$

- New: $(f + df) + (g + dg)$

- Change: $(f + df) + (g + dg) - (f + g) = df + dg$

In my head, I imagine x, the amount you requested, changing silently in a corner. This creates a visible change in f (size df) and g (size dg), and we see the total change as $df + dg$.

It seems we should just combine the total up front, writing $total = 3.5x$ not $total = f(x) + g(x) = 3x + 0.5x$. Normally, we would simplify an equation, but it's sometimes helpful to list every contribution (total = base + shipping + tax). In our case, we see Frank contributes the most to the price.

Remembering the derivative is the "per dx" rate, we write:

$$\frac{d}{dx}\left(f(x) + g(x)\right) = \frac{df}{dx} + \frac{dg}{dx}$$

But ugh, look at all that notation! Let's trim it down:

- Write f instead of $f(x)$. We'll assume a single letter is an entire function, and by the Third Edict of The Grand Math Poombahs, our functions will use a parameter x.

- We'll express the derivative using a single quote (f'), not with a ratio ($\frac{df}{dx}$). We're most interested in the relationship between the parts (addition), not the gritty details of the parts themselves.

So now the addition rule becomes:

$$(f + g)' = f' + g'$$

Much better! Here's how I read it: Take a system made of several parts: $(f + g)$. The change in the overall system, $(f + g)'$, can be found by adding the change from each part.

12.2 Multiplication

Let's try a tricker scenario. Instead of inputs that are added (almost oblivious to each other), what if they are multiplied?

Suppose Frank and George are making a *rectangular* garden for you. Frank handles the width and George takes care of the height. Whenever you clap, they move... but by different amounts!

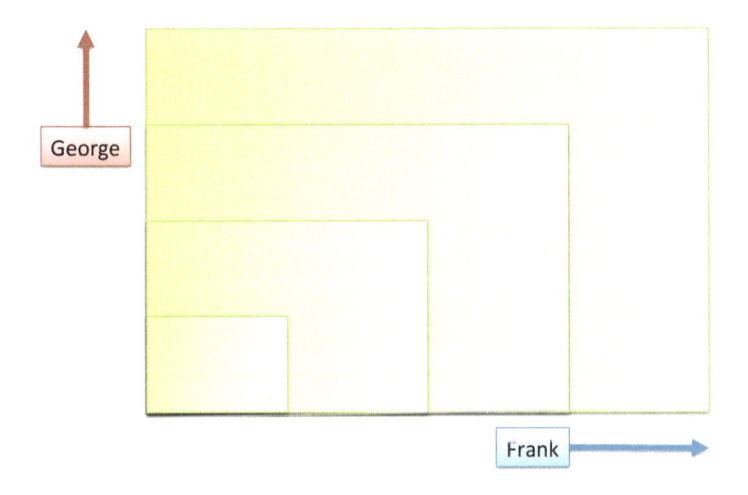

Frank's steps are 3-feet long, but George's are only 2-feet long (zookeeping accident, don't ask). How can we describe the system?

$$Area = width \cdot height = f(x) \cdot g(x)$$
$$f(x) = 3x$$
$$g(x) = 2x$$

We have linear parts, so the derivatives are simple: $f'(x) = 3$ and $g'(x) = 2$. What happens on the next clap?

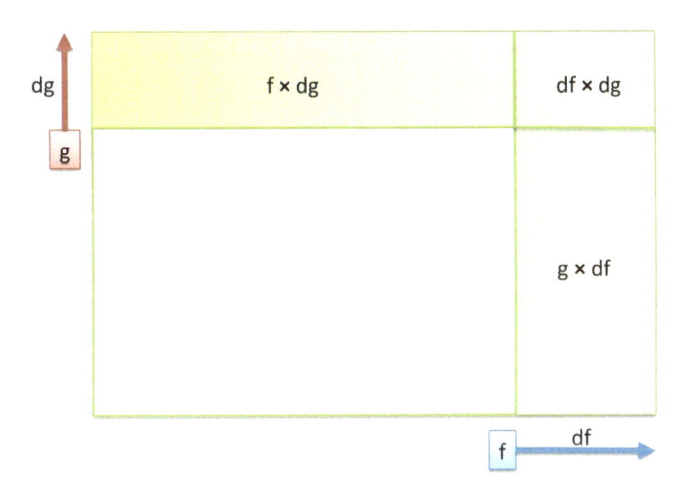

Looks familiar! We have a horizontal strip, a vertical strip, and a corner piece. We can work out the amounts with algebra:

- Original: $f \cdot g$
- New: $(f+df) \cdot (g+dg) = (f \cdot g) + (f \cdot dg) + (g \cdot df) + (df \cdot dg)$
- Change: $f \cdot dg + g \cdot df + df \cdot dg$

Let's see this change more closely:

- The horizontal strip happened when f changed (by df), and g was the same value
- The vertical strip was made when g changed (by dg), and f was the same value
- The corner piece ($df \cdot dg$) happened when the change in one component (df) interacted with the change in the other (dg)

The corner piece is our sample measurement getting tangled on itself, and should be removed. (If we're *forced* to move in whole units, then the corner is fine. But most real-world systems can change continuously, by any decimal number, and we want the measurement artifacts removed.)

To find the total change, we drop the $df \cdot dg$ term (interference between the changes) and get:

$$f \cdot dg + g \cdot df$$

I won't let you forget the derivative is on a "per dx" basis, so we write:

$$\frac{\text{total change}}{dx} = f\frac{dg}{dx} + g\frac{df}{dx}$$

$$(f \cdot g)' = f \cdot g' + g \cdot f'$$

There is an implicit "x" changing off in the distance, which makes f and g move. We hide these details to make the notation simpler.

In English: Take a scenario with multiplied parts. As they change, and continue to be multiplied, add up the new horizontal and vertical strips that are formed.

Let's try out the rule: if we have a 12×8 garden and increment by a whole step, what change will we see?

In this case, we'll use the discrete version of the rule since we're forced to move as a whole step:

- Vertical strip: $f \cdot dg = 12 \cdot 2 = 24$
- Horizontal strip $g \cdot df = 8 \cdot 3 = 24$
- Corner piece: $df \cdot dg = 3 \cdot 2 = 6$
- Total change: $24 + 24 + 6 = 54$

Let's test it. We go from 12×8 (96 square feet) to 15×10 (150 square feet). And yep, the area increase was 150 - 96 = 54 square feet!

12.3 Simple Division (Inverses)

Inverses can be tough to visualize: as x gets bigger, $\frac{1}{x}$ gets smaller. Let's take it slow.

Suppose you're sharing a cake with Frank. You've just cut it in half, about to take a bite and... George shuffles in. He looks upset, and you're not about to mention the fresh set of claw marks.

But you've just cut the cake in half, what can you do?

Cut it again. You and Frank can cut your existing portion in thirds, and give George a piece:

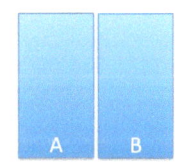

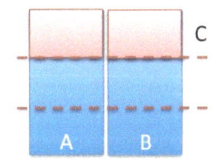

Neat! Now everyone has 1/3 of the total. You gave up 1/3 of your amount (1/2), that is, you each gave George 1/6 of the total.

Time to eat! But just as you're about to bite in... the veggie godfather walks in. Oh, he'll *definitely* want a piece. What do you do?

Cut it again. Everyone smooshes together their portion, cuts it in *fourths*, and hands one piece to the Don. The cake is split evenly again.

This is step-by-step thinking applied to division:

- Your original share is $\frac{1}{x}$ (when x=2, you have 1/2)

- Someone walks in

- Your new share becomes $\frac{1}{x+1}$

How did your amount of cake change? Well, you took your original slice ($\frac{1}{x}$), cut it into the new number of pieces ($\frac{1}{x+1}$), and gave one away (the change is negative):

$$\frac{1}{x} \cdot \frac{-1}{x+1} = \frac{-1}{x(x+1)}$$

We can probably guess that the +1 is a measurement artifact because we forced an integer change in x. If we call the test change dx, we can find the difference between the new amount ($\frac{1}{x+1}$) and the original ($\frac{1}{x}$):

$$\frac{1}{x+dx} - \frac{1}{x} = \frac{x}{x(x+dx)} - \frac{x+dx}{x(x+dx)} = \frac{-dx}{x(x+dx)}$$

After finding the total change (and its annoying algebra), we divide by dx to get the change on a "per dx" basis:

$$\frac{1}{x(x+dx)}$$

Now we remove the leftover dx, the measurement artifact:

CHAPTER 12. THE BASIC ARITHMETIC OF CALCULUS

$$\frac{-1}{x(x+0)} = -\frac{1}{x^2}$$

Phew! We've found how an 1/x split changes as more people are added.

Let's try it out: You are splitting a $1000 bill among 5 people. A sixth person enters, how much money do you save?

You'll personally save 1/5 · 1/6 = 1/30 of the total cost (cut your share into 6 pieces, give the new guy one portion to pay). That's about 3%, or 30. Not bad for a quick calculation!

Let's work it backwards: how large is our group when we're saving about $100 per person? Well, $100 is 1/10 of the total. Since $\frac{1}{3^2} \sim \frac{1}{10}$, we'll hit that savings rate around x=3 people.

And yep, going from 3 to 4 people means each person's share goes from $333.33 to $250 – about $100. Not bad! (If we added people fractionally, we could hit the number exactly.)

12.4 Questions

We didn't explicitly talk about scaling by a constant, such as finding the derivative of $f(x) = 3x$. Can you use the product rule to figure out how it changes? (Hint: imagine a rectangle with a fixed 3 for one side, and x for the other).

Now, how about the addition rule? How would $f(x) = x + x + x$ behave?

CHAPTER 13

FINDING PATTERNS IN THE RULES

We've uncovered the first few rules of calculus:

$$(f+g)' = f' + g'$$
$$(f \cdot g)' = f \cdot g' + g \cdot f'$$
$$\left(\frac{1}{x}\right)' = -\frac{1}{x^2}$$

Instead of blasting through more rules, step back. Is there a pattern here?

13.1 Combining Perspectives

Imagine a business with interacting departments, or a machine with interconnected parts. What happens when we make a change? There's a potential impact on each part.

If our business has 4 departments, and we make a policy change, there are 4 perspectives to consider. It sounds simple when written out: whenever a change happens, see what happens to each part!

No matter the specific interaction between parts F and G (addition, subtraction, multiplication, exponents...), we just have two perspectives to consider:

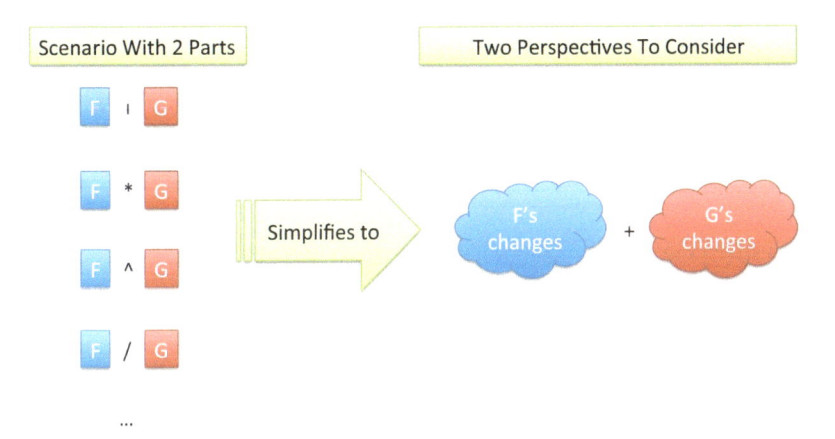

CHAPTER 13. FINDING PATTERNS IN THE RULES

Aha! That's why the rules for $(f+g)'$ and $(f \cdot g)'$ are two added perspectives. The derivative of $\frac{1}{x}$ has a perspective because there's just one "moving part", x. (If you like, there *is* a contribution from "1" about the change it experiences: nothing. No matter how much you yell, 1 stays 1.)

The exact contribution from a perspective depends on the interaction:

- With addition, each part adds a direct change $(df + dg)$.

- With multiplication, each part thinks it'll add a rectangular strip $(f \cdot dg + g \cdot df)$. (I'm using df instead of f' to help us think about the slice being added.)

You might forget the exact form of the multiplication rule. But you can think "The derivative of $f \cdot g$ must be something with df + something with dg."

Let's go further: what about the derivative of $a \cdot b + c$? You guessed it, 3 perspectives that should be added: something involving da plus something involving db plus something involving dc.

We can predict the shape of derivatives for gnarly equations. What's the derivative of:

$$x^y \cdot \frac{u}{v}$$

Wow. I can't rattle that off, but I can say it'll be something involving 4 additions (dx, dy, du and dv). Guess the shape of a derivative, even if you don't know the exact description.

Why does this work? Well, suppose we had a change that was influenced by both da and db, such as $15 \cdot da \cdot db$ – that'd be our instrument interfering with itself!

Only direct changes on a single variable's are counted (such as $3da$ or $12db$), and "changes on changes" like $15 \cdot da \cdot db$ are ignored.

13.2 Dimensional Intuition

Remember that derivatives are a fancier form of division. What happens when we do a division like $\frac{x^3}{x}$? We divide volume by length, and get area (one dimension down).

What happens when we do $\frac{d}{dx}x^3$? You might not know yet, but you can bet we'll be dropping a dimension.

Dimensions	Example	Description
3	x^3	Volume (cubic growth)
2	x^2	Area (square growth)
1	x	Length (linear growth)
0	c	Constant (no change)
-1	$\frac{1}{x}$	Inverse length ("per length")
-2	$\frac{1}{x^2}$	Inverse area ("per area")
-3	$\frac{1}{x^3}$	Inverse volume ("per volume")

When we divide or take a derivative, we drop a dimension and hop down the table. Volume is built from area (slices dx thick), area is built from length (dx wide).

A constant value, like 3, has no dimension in the following sense: it never produces a set of slices. There will never be a jump to the next value. Once we have a constant value, we get "stuck" on the table because the 0 pattern always produces another 0 pattern (volume to area to lines to constant to zero to zero...).

We can have negative dimensions as well: "per length", "per area", "per volume", etc. Derivatives still decrease the dimension, so when seeing $\frac{1}{x}$, we know the derivative will resemble $\frac{1}{x^2}$ as the dimension drops a level.

An important caveat: Calculus only cares about quantities, not their dimension. The equations will happily combine x and x^2, even though we know you can't mix length with area. Units add a level of meaning from *outside* the equation, that help keep things organized and warn us if we've gone awry (if you differentiate area and get volume, you know something went wrong).

13.3 Thinking With Dimensions

You can think about the dimension of a derivative without digging into a specific formula. Imagine the following scenario:

- Take a string and wrap it tight around a quarter. Take another string and wrap it tight around the Earth.

- Lengthen both strings, adding more to the end, so there's a 1-inch gap all the way around the quarter, and a 1-inch gap all the way around the Earth (like having a ring floating around Saturn).

- Quiz: Which scenario needed more extra string to create? Is it more string to put a 1-inch gap around the quarter, or a 1-inch gap around the Earth?

We can crunch through the formula, but think higher-level. Because circumference is a 1-d line, its reaction to change (the derivative) will be a constant. No matter the current size, the circumference will change the same amount with every push. So, the extra string needed is the same for both! (About 6.28 inches per inch of radius increase).

Now, suppose we were painting the sphere instead of putting a string around it. Ah, well, area is squared, therefore the derivative is a dimension lower (linear). If we have a 5-inch and 10-inch sphere, and make them 1-inch bigger each, the larger sphere will require double the *extra* paint.

13.4 Questions

Let's think about the derivative of x^3, a growing cube.

1) What dimension should the derivative of x^3 have?
2) How many viewpoints should $x^3 = x \cdot x \cdot x$ involve?
3) Have a guess for the derivative? Does it match with how you'd imagine a cube to grow?

CHAPTER 14

THE FANCY ARITHMETIC OF CALCULUS

Here's the rules we have so far:

$$(f+g)' = f'+g'$$
$$(f \cdot g)' = f \cdot g' + g \cdot f'$$
$$\left(\frac{1}{x}\right)' = -\frac{1}{x^2}$$

Let's add a few more to our collection.

14.1 Power Rule

We've worked out that $\frac{d}{dx}x^2 = 2x$:

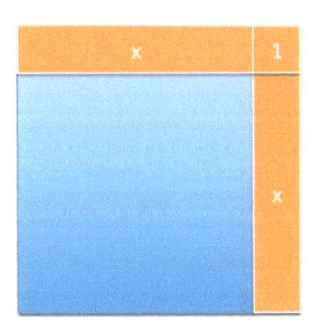

We can visualize the change, and ignore the artificial corner piece. Now, how about visualizing x^3?

CHAPTER 14. THE FANCY ARITHMETIC OF CALCULUS

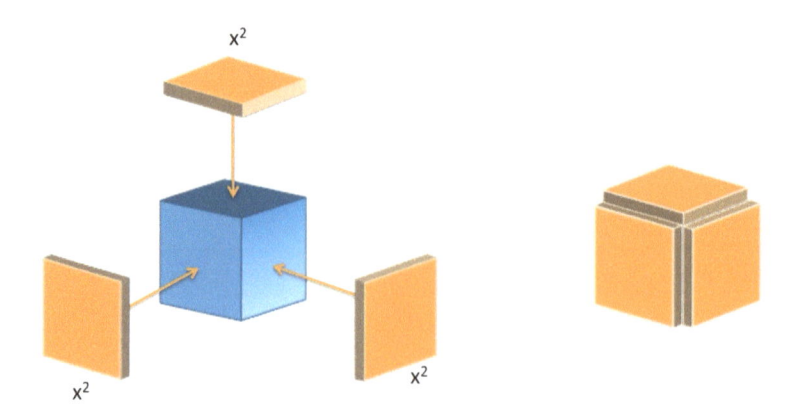

The process is similar. We can glue a plate to each side to expand the cube. The "missing gutters" represent artifacts, where our new plates would interact with each other.

I have to keep reminding myself: the gutters aren't real! They represent growth that doesn't happen at this step. After our growth, we "melt" the cube into its new, total area, and grow again. Counting the gutters would overestimate the growth that happened in this step. (Now, if we're forced to take integer-sized steps, then the gutters are needed – but with infinitely-divisible decimals, we can change smoothly.)

From the diagram, we might guess:

$$\frac{d}{dx}x^3 = 3x^2$$

And that's right! But we had to visualize the result. Abstractions like algebra let us handle scenarios we can't visualize, like a 10-dimensional shape. Geometric shapes are a nice, visual starting point, but we need to move beyond them.

We might begin analyzing a cube with using algebra like this:

$$(x+dx)^3 = (x+dx)(x+dx)(x+dx) = (x^2 + 2x \cdot dx + (dx)^2)(x+dx) = ...$$

Yikes. The number of terms is getting scary, fast. What if we wanted the 10th power? Sure, there are algebra shortcuts, but let's think about the problem holistically.

Our cube $x^3 = x \cdot x \cdot x$ has 3 components: the sides. Call them a, b and c to keep 'em straight. Intuitively, we know the total change has a contribution from each side:

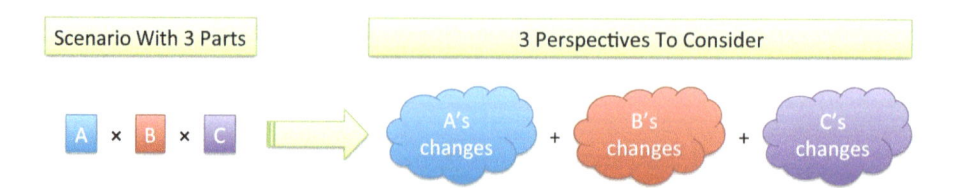

What change does each side think it's contributing?

- **a thinks:** My change (da) is combined with the other, unmoving, sides ($b \cdot c$) to get $da \cdot b \cdot c$
- **b thinks**: My change (db) is combined with the other sides to get $db \cdot a \cdot c$
- **c thinks:** My change (dc) is combined with the other sides to get $dc \cdot a \cdot b$

Each change happens separately, and there's no "crosstalk" between da, db and dc (such crosstalk leads to gutters, which we want to ignore). The total change is:

$$A's\ changes + B's\ changes + C's\ changes = (da \cdot b \cdot c) + (db \cdot a \cdot c) + (dc \cdot a \cdot b)$$

Let's write this in terms of x, the original side. Every side is identical, ($a = b = c = x$) and the changes are the same ($da = db = dc = dx$), so we get:

$$(dx \cdot x \cdot x) + (dx \cdot x \cdot x) + (dx \cdot x \cdot x) = x^2 \cdot dx + x^2 \cdot dx + x^2 \cdot dx = 3x^2 \cdot dx$$

Converting this to a "per dx" rate we have:

$$\frac{d}{dx} x^3 = 3x^2$$

Neat! Now, the brain-dead memorization strategy is to think "Pull down the exponent and decrease it by one". That isn't learning!

Think like this:

- x^3 has 3 identical perspectives.
- When the system changes, all 3 perspectives contribute identically. Therefore, the derivative will be $3 \cdot something$.
- The "something" is the change in one side (dx) multiplied by the remaining sides ($x \cdot x$). The changing side goes from x to dx and the exponent lowers by one.

We can reason through the rule! For example, what's the derivative of x^5?

Well, it's 5 identical perspectives ($5 \cdot something$). Each perspective is me changing (dx) and the 4 other guys staying the same ($x \cdot x \cdot x \cdot x = x^4$). So the combined perspective is just $5x^4$.

The general Power Rule:

$$\frac{d}{dx} x^n = nx^{n-1}$$

Now we can memorize the shortcut "bring down the exponent and subtract", just like we know that putting a 0 after a number multiplies by 10. Shortcuts are fine once you know *why* they work!

14.2 Integrals of Powers

Let's try integrating a power, reverse engineering a set of changes into the original pattern.

Imagine a construction site. Day 1, they order three 1×1 wooden planks. The next day, they order three 2×2 wooden planks. Then three 3×3 planks. Then three 4×4 planks. What are they building?

My guess is a cube. They are building a shell, layer by layer, and perhaps putting grout between the gutters to glue them together.

Similarly, if we see a series of changes like $3x^2$, we can visualize the plates being assembled to build a cube:

$$\int 3x^2 = x^3$$

Ok – we took the previous result and worked backward. But what about the integral of plain old x^2? Well, just imagine that incoming change is being split 3 ways:

$$x^2 = \frac{x^2}{3} + \frac{x^2}{3} + \frac{x^2}{3} = \frac{1}{3}3x^2$$

Ah! Now we have 3 plates (each 1/3 of the original size) and we can integrate a smaller cube. Imagine the "incoming material" being split into 3 piles to build up the sides:

$$\int x^2 = \int \frac{1}{3} \cdot 3x^2 = \frac{1}{3} \int 3x^2 = \frac{1}{3} x^3$$

If we have 3 piles of size x^2, we can make a full-sized cube. Otherwise, we build a mini-cube, 1/3 as large.

The general integration rule is:

$$\int x^n = \frac{1}{n+1} x^{n+1}$$

After some practice, you'll do the division automatically. But now you know *why* it's needed: we have to split the incoming "change material" among several sides. (Building a square? Share changes among 2 sides. Building a cube? Share among 3 sides. Building a 4d hypercube? Call me.)

14.3 The Quotient Rule

We've seen the derivative of an inverse (a "simple division"):

$$\frac{d}{dx}\frac{1}{x} = -\frac{1}{x^2}$$

Remember the cake metaphor? We cut our existing portion ($\frac{1}{x}$) into x slices, and give one away.

Now, how can we find the derivative of $\frac{f}{g}$? One component in the system is trying to grow us, while the other divides us up. Which wins?

Abstraction to the rescue. When finding the derivative of x^3, we imagined it as $x^3 = a \cdot b \cdot c$, which helped simplify the interactions. Instead of a mishmash of x's being multiplied, it was just 3 distinct perspectives to consider.

Similarly, we can rewrite $\frac{f}{g}$ as two perspectives:

$$\frac{f}{g} = a \cdot b$$

We know $a = f$ and $b = \frac{1}{g}$. From this zoomed-out view, it looks like a normal, rectangular, product-rule scenario:

$$(a \cdot b)' = da \cdot b + db \cdot a$$

It's our little secret that b is really $\frac{1}{g}$, which behaves like a division. We just want to think about the big picture of how the rectangle changes.

Now, since a is just a rename of f, we can swap in $da = df$. But how do we swap out b? Well, we have:

$$b = \frac{1}{g}$$

$$\frac{db}{dg} = -\frac{1}{g^2}$$

$$db = -\frac{1}{g^2} dg$$

Ah! This is our cake cutting. As g grows, we lose $db = -\frac{1}{g^2} dg$ from the b side. The total impact is:

$$(a \cdot b)' = (da \cdot b) + (db \cdot a) = \left(df \cdot \frac{1}{g}\right) + \left(\frac{-1}{g^2} dg \cdot f\right)$$

This formula started with the product rule, and we plugged in their real values. Might as well put f and g back into $(a \cdot b)'$, to get the Quotient Rule (aka the Division Rule):

$$\left(\frac{f}{g}\right)' = \left(df \cdot \frac{1}{g}\right) + \left(\frac{-1}{g^2} dg \cdot f\right)$$

Many textbooks re-arrange this relationship, like so:

$$\left(df \cdot \frac{1}{g}\right) + \left(\frac{-1}{g^2} dg \cdot f\right) = \frac{g \cdot df}{g^2} - \frac{f \cdot dg}{g^2} = \frac{g \cdot df - f \cdot dg}{g^2}$$

And I don't like it, no ma'am, not one bit! This version no longer resembles its ancestor, the product rule.

In practice, the Quotient Rule is a torture device designed to test your memorization skills; I rarely remember it. Just think of $\frac{f}{g}$ as $f \cdot \frac{1}{g}$, and use the product rule like we've done.

14.4 Questions

Let's do a few warm-ups to test our skills. Can you solve these bad boys?

$$\frac{d}{dx}x^4 = ?$$

$$\frac{d}{dx}3x^5 = ?$$

(You can check your answers with Wolfram Alpha, such as `d/dx x^^4`.)

Again, don't get lost in the symbols. Think "I have x^4 – what pattern of changes will I see as I make x larger?".

Ok! How about working backwards, and doing some integrals?

$$\int 2x^2 = ?$$

$$\int x^3 = ?$$

Ask yourself, "What original pattern would create steps in the pattern $2x^2$?"

Trial-and-error is ok! Try a formula, test it, and adjust it. Personally, I like to move aside the 2 and just worry about the integral of x^2:

$$\int 2x^2 = 2\int x^2 = ?$$

How do you know if you're right? Take the derivative – *you* are the antiques dealer! I brought you a pattern of shards ($2x^2$) and you need to tell me the vase they came from. Once you have guessed a vase, break a replica in the back room and make sure you get $2x^2$ back out. Then you'll be confident in your answer (*and your manager will be thrilled!*).

We're getting ready to work through the circle equations ourselves, and recreate results found by Archimedes, likely the greatest mathematician of all time.

CHAPTER 15

DISCOVERING ARCHIMEDES' FORMULAS

In the preceding lessons we uncovered a few calculus relationships, the "arithmetic" of how systems change:

Interaction	Overall Change
Addition	$(f + g)' = f' + g'$
Multiplication	$(f \cdot g)' = f \cdot dg + g \cdot df$
Powers	$(x^n)' = \dfrac{d}{dx}x^n = nx^{n-1}$
Inverse	$\left(\dfrac{1}{x}\right)' = -\dfrac{1}{x^2}$
Division	$\left(\dfrac{f}{g}\right)' = \left(df \cdot \dfrac{1}{g}\right) + \left(\dfrac{-1}{g^2}dg \cdot f\right)$

How do these rules help us?

- If we have an existing equation, the rules are a shortcut to finding the step-by-step pattern. Instead of visualizing a growing square, or cube, the Power Rule lets us crank through the derivatives of x^2 and x^3. Whether x^2 refers to a literal square or just the multiplication $x \cdot x$ isn't important – we'll get the pattern of changes.

- If we have a set of changes, the rules help us reverse-engineer the original pattern. Getting changes like $2x$ or $14x$ is a hint that *something* $\cdot x^2$ was the original pattern.

Learning to think with Calculus means we can use X-Ray and Time-lapse vision to imagine changes taking place, and use the rules to work out the specifics. Eventually, we might not visualize anything, and just work with the symbols directly (as you likely do with arithmetic today).

In the start of the course, we morphed a ring into a circle, then a sphere, then a shell:

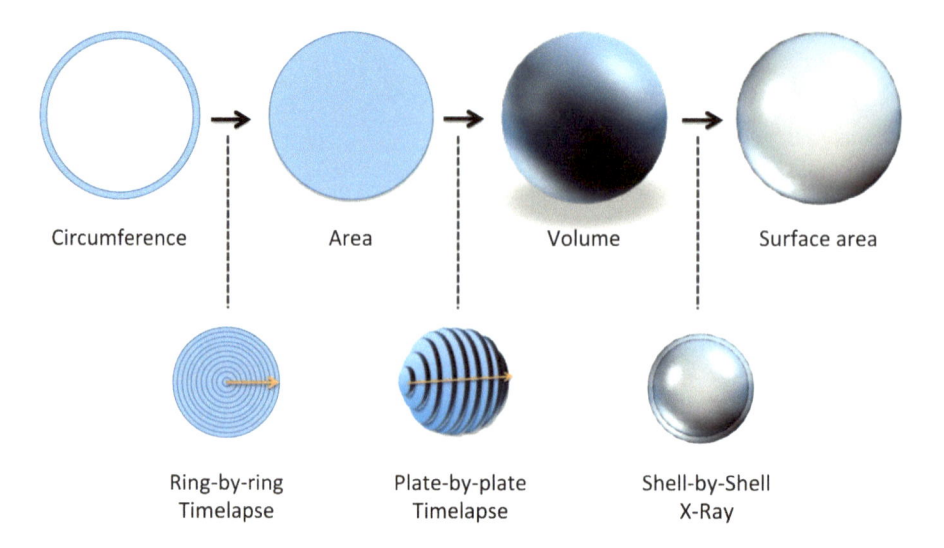

Chapter 15. Discovering Archimedes' Formulas

| Circumference | Area | Volume | Surface area |

| Ring-by-ring Timelapse | Plate-by-plate Timelapse | Shell-by-Shell X-Ray |

With the official rules in hand, we can blast through the calculations and find the circle/sphere formulas on our own. It may sound strange, but the formulas feel different to me – almost alive – when you see them morphing in front of you. Let's jump in.

15.1 Changing Circumference To Area

Our first example of "step-by-step" thinking was gluing a sequence of rings to make a circle:

Symbolic Description	Solution	Notes
$area = \int_0^r 2\pi r \, dr$	$= 2\pi \int_0^r r \, dr$ $= 2\pi \left(\frac{1}{2}r^2\right)$ $= \pi r^2$	Work backwards to the integral. If $\frac{d}{dx}r^2 = 2r$ that means $\frac{d}{dx}\frac{1}{2}r^2 = r$

When we started, we needed a lot of visualization. We had to unroll the rings, line them up, realize they made a triangle, then use $\frac{1}{2} base \cdot height$ to get the area. Visual, tedious... and necessary. We need to feel what's happening before working with raw equations.

Here's the symbolic approach:

Chapter 15. Discovering Archimedes' Formulas

Symbolic Description	Solution	Notes
$area = \int_0^r 2\pi r\, dr$	$= 2\pi \int_0^r r\, dr$ $= 2\pi \left(\frac{1}{2}r^2\right)$ $= \pi r^2$	Work backwards to the integral. If $\frac{d}{dx}r^2 = 2r$ that means $\frac{d}{dx}\frac{1}{2}r^2 = r$

Let's walk through it. The notion of a "ring-by-ring timelapse" sharpens into "integrate the rings, from nothing to the full radius" and ultimately:

$$\text{Area} = \int_0^r 2\pi r\, dr$$

Each ring has height $2\pi r$ and width dr, and we want to accumulate that area to make our disc.

How can we solve this equation? By working backwards. We can move the 2π part outside the integral (remember the scaling property?) and focus on the integral of r:

$$2\pi \int_0^r r\, dr = ?$$

What pattern makes steps of size r? Well, we know that r^2 creates steps of size $2r$, which is twice what we need. Half that should be perfect. Let's try it out:

$$\frac{d}{dr}\frac{1}{2}r^2 = \frac{1}{2}\frac{d}{dr}r^2 = \frac{1}{2}2r = r$$

Yep, $\frac{1}{2}r^2$ gives us the steps we need! Now we can plug in the solution to the integral:

$$\text{Area} = 2\pi \int_0^r r\, dr = 2\pi \frac{1}{2}r^2 = \pi r^2$$

This is the same result as making the ring-triangle in the first lesson, but we manipulated equations, not diagrams. Not bad! It'll help even more once we get to 3d...

15.2 Changing Area To Volume

Let's get fancier. We can take our discs, thicken them into plates, and build a sphere:

CHAPTER 15. DISCOVERING ARCHIMEDES' FORMULAS

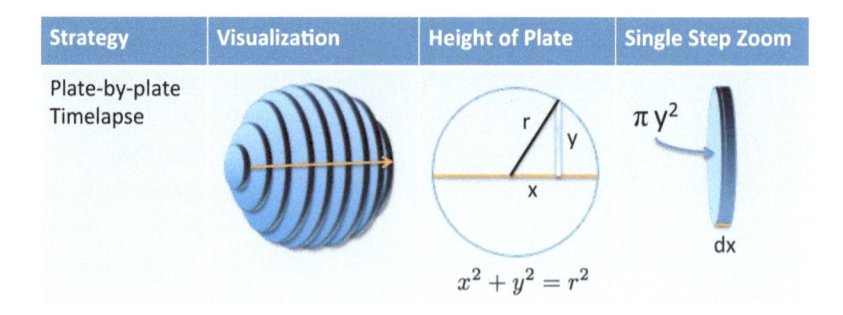

Strategy	Visualization	Height of Plate	Single Step Zoom
Plate-by-plate Timelapse		$x^2 + y^2 = r^2$	πy^2 dx

Let's walk slowly. We have several plates, each at a different "x-coordinate". What's the size of a single plate?

The plate has a thickness (dx), and its own radius. The radius of the plate is its height from the x-axis, which we can call y.

It's a little confusing at first: r is the radius of the entire sphere, but y is the (usually smaller) radius of an individual plate under examination. In fact, only the center plate ($x = 0$) will have its radius the same as the entire sphere. The "end plates" don't have a height at all.

And by the Pythagorean theorem, we have a connection between the x-position of the plate, and its height (y):

$$x^2 + y^2 = r^2$$

Ok. We have size of each plate, and can integrate to find the volume, right?

Not so fast. Instead of starting on the left side, with a negative x-coordinate, moving to 0, and then up to the max, let's just think about a sphere as two halves:

To find the total volume, get the volume of one half, and double it. This is a common trick: if a shape is symmetrical, get the size of one part and scale it up. Often, it's easier to work out "0 to max" than "min to max", especially when "min" is negative.

Ok. *Now* let's solve it:

CHAPTER 15. DISCOVERING ARCHIMEDES' FORMULAS

Symbolic	Solution	Notes
$Volume = 2\int_0^r \pi y^2 dx$	$= 2\int_0^r \pi y^2 \, dx$	Write height in terms of x
		$x^2 + y^2 = r^2$
	$= 2\int_0^r \pi(\sqrt{r^2 - x^2})^2 \, dx$	$y = \sqrt{r^2 - x^2}$
	$= 2\pi \int_0^r r^2 - x^2 \, dx$	Work backwards to find integrals
	$= 2\pi \left((r^2)x - \frac{1}{3}x^3\right)$	
	$= 2\pi \left((r^2)r - \frac{1}{3}r^3\right)$	Find volume at full radius (x=r)
	$= 2\pi \left(\frac{2}{3}r^3\right)$	
	$= \frac{4}{3}\pi r^3$	

Whoa! Quite an equation, there. It seems like a lot, but we'll work through it:

$$Volume = 2\int_0^r \pi y^2 \, dx$$

First off, three variables (r, y, x) is too many to have flying around in a single equation. We'll write the height of each plate (y), in terms of the others:

$$height = y = \sqrt{r^2 - x^2}$$

The square root looks intimidating at first, but it's being plugged into y^2 and the exponent will cancel it out. After plugging in y and moving π outside the integral, we have the much nicer:

$$Volume = 2\int_0^r \pi \left(\sqrt{r^2 - x^2}\right)^2 \, dx$$

$$Volume = 2\pi \int_0^r r^2 - x^2 \, dx$$

The parentheses are often dropped because it's understood that dx is multiplied by the entire size of the step. We know the step is $(r^2 - x^2)dx$ and not $r^2 - (x^2 dx)$.

Let's talk about r and x for a minute. r is the radius of the entire *sphere*, such as "15 inches". You can imagine asking "I want the volume of a sphere with a radius of 15 inches". Fine.

To figure this out, we'll create plates at each x-coordinate, from $x = 0$ up to $x = 15$ (and double it). x is the bookkeeping entry that remembers which plate we're on. We could work out the volume from $x = 0$ to $x = 7.5$, let's say, and we'd build a partial sphere (maybe useful, maybe not). But we want the whole shebang, so we let x go from 0 to the full r.

Time to solve this bad boy. What equation has steps like $r^2 - x^2$?

CHAPTER 15. DISCOVERING ARCHIMEDES' FORMULAS

First, let's use the addition rule: steps like $a - b$ are made from two patterns (one making a, the other making b).

Let's look at the first pattern, the steps of size r^2. We're moving along the x-axis, and r is a number that never changes: it's 15 inches, the size of our sphere. This max radius never depends on x, the position of the current plate.

When a scaling factor doesn't change during the integral (r, π, etc.), it can be moved outside and scaled up at the end. So we get:

$$\int r^2 \, dx = r^2 \int dx = r^2 x$$

In other words, $r^2 \cdot x$ is a linear trajectory that contributes a constant r^2 at each step.

Cool. How about the integral of $-x^2$? First, we can move out the negative sign and take the integral of x^2:

$$-\int x^2 \, dx = ?$$

We've seen this before. Since x^3 has steps of $3x^2$, taking $1/3$ of that amount ($\frac{x^3}{3}$) should be just right. And we can check that our integral is correct:

$$\frac{d}{dx}\left(-\frac{1}{3}x^3\right) = -\frac{1}{3}\frac{d}{dx}x^3 = -\frac{1}{3}3x^2 = -x^2$$

It works out! Over time, you'll learn to trust the integrals you reverse-engineer, but when starting out, it's good to check the derivative. With the integrals solved, we plug them in:

$$2\pi \int r^2 - x^2 \, dx = 2\pi(r^2 x - \frac{1}{3}x^3)$$

What's left? Well, our formula still has x inside, which measures the volume from 0 to some final value of x. In this case, we want the full radius, so we set $x = r$:

$$2\pi(r^2 x - \frac{1}{3}x^3) \xrightarrow[\text{set } x=r]{} 2\pi\left((r^2)r - \frac{1}{3}r^3\right) = 2\pi\left(r^3 - \frac{1}{3}r^3\right) = 2\pi\frac{2}{3}r^3 = \frac{4}{3}\pi r^3$$

Tada! You've found the volume of a sphere (or another portion of a sphere, if you use a different range for x).

Think that was hard work? You have no idea. That one-line computation took Archimedes, one of the greatest geniuses of all time, tremendous effort to figure out. He had to imagine some spheres, and a cylinder, and some cones, and a fulcrum, and imagine them balancing and... let's just say when he found the formula, he had it written on his grave. Your current intuition would have saved him incredible effort (see this video).

15.3 Changing Volume To Surface Area

Now that we have volume, finding surface area is much easier. We can take a thin "peel" of the sphere with a shell-by-shell X-Ray:

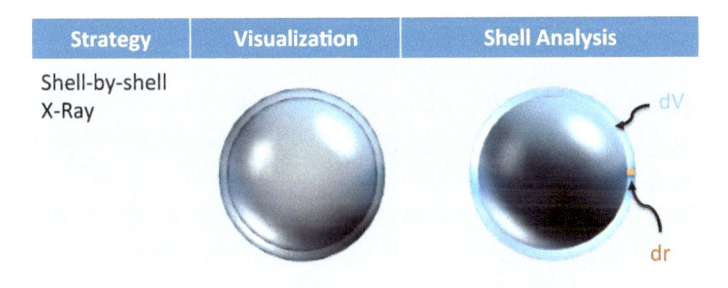

Strategy	Visualization	Shell Analysis
Shell-by-shell X-Ray		dV, dr

I imagine the entire shell as "powder" on the surface of the existing sphere. How much powder is there? It's dV, the change in volume. Ok, what is the *area* the powder covers?

Hrm. Think of a similar question: how much area will a bag of mulch cover? Get the volume, divide by the desired thickness, and you have the area covered. If I give you 300 cubic inches of dirt, and spread it in a pile 2 inches thick, the pile will cover 150 square inches. After all, if $Area \cdot Thickness = Volume$ then $Area = \frac{Volume}{Thickness}$.

In our case, dV is the volume of the shell, and dr is its thickness. We can spread dV along the thickness we're considering (dr) and see how much area we added: $\frac{dV}{dr}$, the derivative.

This is where the right notation comes in handy. We can think of the derivative as an abstract, instantaneous rate of change (V'), or as a specific ratio ($\frac{dV}{dr}$). In this case, we want to consider the individual elements, and how they interact (volume of shell / thickness of shell).

So, given the relation,

$$\text{Area of shell} = \frac{\text{Volume of shell}}{\text{Depth of shell}} = \frac{dV}{dr}$$

we figure out:

$$\frac{d}{dr}\text{Volume} = \frac{d}{dr}\frac{4}{3}\pi r^3 = \frac{4}{3}\pi \frac{d}{dr}r^3 = \frac{4}{3}\pi(3r^2) = 4\pi r^2$$

Wow, that was fast! The order of our morph (Circumference → Area → Volume → Surface area) made the last step simple. We could try to spin a circumference into surface area directly, but it's more complex.

As we cranked through this formula, we "dropped the exponent" on r^3 to get $3r^2$. Remember the total change comes from 3 perspectives that contribute an equal share: $\frac{d}{dr}r^3 = r^2 + r^2 + r^2 = 3r^2$.

15.4 2000 Years Of Math In A Day

The steps we worked through took 2000 years of thought to discover, by the greatest geniuses no less. Calculus is such a broad and breathtaking viewpoint that it's difficult to imagine where it *doesn't* apply. It's just about using X-Ray and Time-Lapse vision:

- **Break things down.** In your current situation, what's the next thing that will happen? And after that? Is there a pattern here? (Getting bigger, smaller, staying the same.) Is that knowledge useful to you?

- **Find the source.** You're seeing a bunch of changes – what caused them? If you know the source, can you predict the end-result of all the changes? Is that prediction helpful?

We're used to analyzing equations, but I hope it doesn't stop there. Numbers can describe mood, spiciness, and customer satisfaction; step-by-step thinking can describe battle plans and psychological treatment. Equations and geometry are just nice starting points to analyze. Math isn't about equations, and music isn't about sheet music – they point to the idea inside the notation.

While there are more details for other derivatives, integration techniques, and how infinity works, you don't need them to start thinking with Calculus. What you discovered today would have brought a tear to Archimedes' eye, and that's a good enough start for me.

Happy math.

Afterword

In martial arts, a black belt doesn't indicate mastery. It means you're a competent-enough beginner who can *now* start learning.

By now, we have a solid intuition for Calculus: it explores patterns with X-Ray and Time-Lapse viewpoints, shows tradeoffs in how objects are made, and gives us improved multiplication and division. (These insights are more than I had after years of classes.)

From here, the path is up to you. An intuitive appreciation is wonderful, and if you wish to sharpen your understanding, follow the guide in the appendix.

If a curious friend asks *What is Calculus all about?* and you look forward to answering, I've done my job.

Unroll the Rings

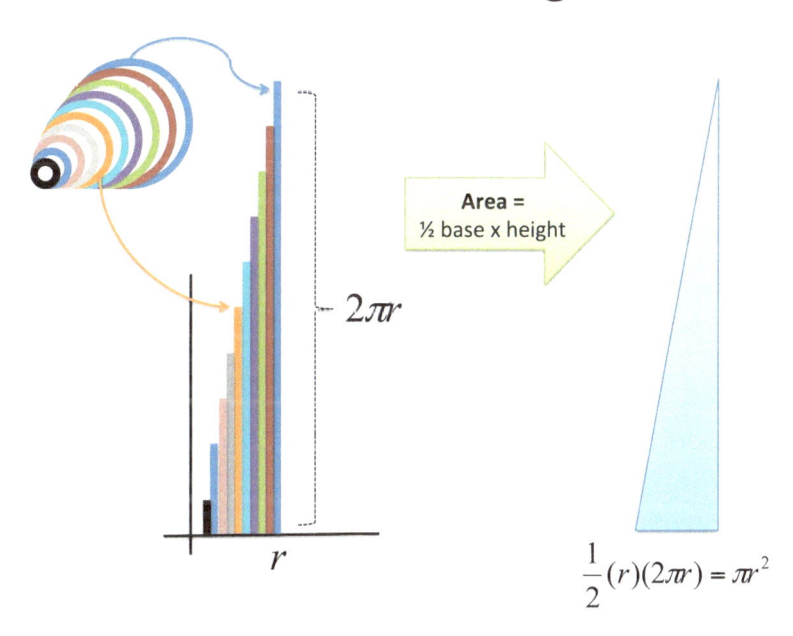

Area = ½ base x height

$2\pi r$

$\frac{1}{2}(r)(2\pi r) = \pi r^2$

Happy math.

Keep In Touch

If you'd like clear, simple insights for Calculus and other math topics, join the BetterExplained newsletter:

http://betterexplained.com/newsletter

About the Author

Kalid Azad is the founder of `http://betterexplained.com`, a teaching website that helps millions of students annually. Its lessons have appeared in hundreds of university and high school courses, Science Magazine, and the blogs for the New York Times, The Atlantic, Scientific American, and the National Academy of Sciences. His book *Math, Better Explained* is a well-received Amazon bestseller.

While studying computer science at Princeton University, Kalid became enamored with finding the *Aha!* moment that demystified a confusing topic. He has been teaching and writing professionally for over a decade, including chapters in the best-selling "How to Program" textbooks (from Deitel, Inc.) and technical whitepapers for Microsoft.

APPENDIX: LEARNING CHECKLIST

Check your Calculus intuition and skills using the questions below.

Intuitive Appreciation (Chapters 1-3)

Describe, in your own words:

- What Calculus does
- X-Ray Vision
- Time-lapse Vision
- The tradeoffs when splitting a circle into rings, wedges, or boards
- How to build a 3d shape from 2d parts

Technical Description (Chapters 4-5)

Describe, in your own words:

- Integral
- Derivative
- Integrand (a single step)
- Bounds of integration

Skills:

- Describe a Calculus action (splitting a circle into rings) using the official language
- Enter the official language into Wolfram Alpha to solve the problem

Theory I (Chapters 6-8)

Describe, in your own words:

- How integrals/derivatives relate to multiplication/division

Skills:

- Find the derivative/integral of a line
- Find the derivative/integral of a constant
- Find the derivative/integral of a square
- Recognize the common notations for the derivative
- Estimate the change in $f(x) = x^2$ using a step of size dx

APPENDIX: LEARNING CHECKLIST

Theory II (Chapters 9-14)

Describe, in your own words:

- How an infinite process can have a finite result
- How a process with limited precision can point to a perfect result
- The formal definition of the derivative
- Estimate the change in $f(x) = x^2$ using a step of size dx, and let dx go to zero. Verify the limit using Wolfram Alpha.
- The Fundamental Theorem of Calculus (FTOC)

Derive and put into your own words:

- The addition rule: $(f + g)' = ?$
- The product rule: $(f * g)' = ?$
- The inverse rule: $(\frac{1}{x})' = ?$
- The power rule: $(x^n)' = ?$
- The quotient rule: $(\frac{f}{g})' = ?$
- Solve $\frac{d}{dx} 3x^5$ on your own and verify with Wolfram Alpha
- Solve $\int 2x^2$ on your own and verify with Wolfram Alpha

Performance (Chapter 15)

Describe how to turn the circumference of a circle into the area of a circle:

- Explain your plan in plain English
- Explain your plan using the official math notation
- Apply the rules of Calculus to your equation and calculate the result
- Verify the result using Wolfram Alpha
- Repeat the steps above, turning the area of a circle into the volume of a sphere
- Repeat the steps above, turning the volume of a sphere into the surface area of a sphere

APPENDIX: CALCULUS STUDY PLAN

Week 1

Read *Calculus, Better Explained*:

- Day 1 - Intuitive Appreciation (Chapters 1-3)
- Day 2 - Technical Description (Chapters 4-5)
- Day 3 - Theory I (Chapters 6-8)
- Day 4 - Theory II (Chapters 9-14)
- Day 5 - Performance (Chapter 15)

You don't need to memorize every result; follow the learning checklist and use online tools to help answer questions if you get stuck.

Weeks 2-12

Begin a traditional Calculus course[1], such as:

- *Elementary Calculus: An Infinitesimal Approach* by Jerome Keisler (2002). This book is based on infinitesimals (an alternative to limits, which I like) and has plenty of practice problems. Available in print or free online.

- *Calculus Made Easy* by Silvanus Thompson (1914). This book follows the traditional limit approach, and is written in a down-to-earth style. Available on Project Gutenberg and print.

- MIT 1801: Single Variable Calculus. Includes video lectures, assignments, exams, and solutions. Available free online.

As you go through the traditional course, keep this in mind.

- **Review the intuitive definition.** Rephrase technical definitions in terms that make sense to you.

- **It's completely fine to use online tools for help.** When stuck, get a hint, fix your mistakes, and try solving a new problem on your own.

- **Relate graphs back to shapes.** Most courses emphasize graphs and slopes; convert the concepts to shapes to help visualize them.

- **Skip limits if you get stuck.** Limits (and infinitesimals) were invented after the majority of Calculus. If you struggle, move on and return later.

Enjoy.

---

[1]Visit http://betterexplained.com/calculus/book for clickable URLs.